SpringerBriefs in Applied Sciences and Technology

Thermal Engineering and Applied Science

Series Editor

Francis A. Kulacki, University of Minnesota, USA

For further volumes:
http://www.springer.com/series/10305

Pradipta Kumar Panigrahi
Krishnamurthy Muralidhar

Schlieren and Shadowgraph Methods in Heat and Mass Transfer

 Springer

Pradipta Kumar Panigrahi
Department of Mechanical Engineering
Indian Institute of Technology Kanpur
Kanpur, Uttar Pradesh
India

Krishnamurthy Muralidhar
Department of Mechanical Engineering
Indian Institute of Technology Kanpur
Kanpur, Uttar Pradesh
India

ISSN 2191-530X   ISSN 2191-5318 (electronic)
ISBN 978-1-4614-4534-0   ISBN 978-1-4614-4535-7 (eBook)
DOI 10.1007/978-1-4614-4535-7
Springer New York Heidelberg Dordrecht London

Library of Congress Control Number: 2012941279

Printed on acid-free paper

Springer is part of Springer Science+Business Media (www.springer.com)

Preface

Careful and continuous detection of velocity and changes in temperature over a distributed region is required in quite a few contexts. Using appropriate light sources, it is possible to map velocity and temperature fields over a cross-section and as a function of time. If the region of interest is transparent, refractive index would be a field variable and beam bending effects can be used to extract information about temperature. Similar principles can be used to determine concentration of salts dissolved in liquids. Time-lapsed images of light intensity can also be used to determine fluid velocity. The fundamentals involved in such measurements, experimental setup, and data analyses constitute the scope of the monograph.

Over the past three decades, laser measurement techniques have become popular, though primarily as a flow visualization tool. Literature in the past decade, however, has emphasized the possibility of quantitative measurements. Optical imaging can be carried out using transmitted light in transparent media or scattered light from particles carried by the flow. Optical methods that utilize the dependence of refractive index of light on density (and indirectly on concentration and temperature) can be configured in many different ways. Three available routes considered are interferometry, schlieren, and shadowgraph. Images recorded in these configurations can be analyzed to yield time sequences of 3D distributions of the transported variables.

Optical methods are non-intrusive, inertia free, and can image cross-sections of the experimental apparatus. With the experiment suitably carried out, 3D physical domains with unsteady processes can be accommodated. Thus, optical methods promise to breach the holy grail of measurements by extracting unsteady 3D data in applications related to transport phenomena. The monograph is intended to provide the readers a glimpse of this exciting development.

A companion volume by the authors entitled *Imaging Heat and Mass Transfer Processes—Visualization and Analysis* is concerned with refractive index-based imaging in a variety of applications.

Pradipta K. Panigrahi
Krishnamurthy Muralidhar

Acknowledgments

We are thankful to quite a few doctoral and Master's students who worked with us on laser imaging of fluid and thermal systems. We would like to acknowledge the contributions of the following doctoral students who have graduated and are pursuing laser measurements as a full-time career.

1. Sunil Punjabi
2. Atul Srivastava
3. Sunil Verma
4. Surendra K. Singh
5. Anamika S. Gupta

We have drawn material from their doctoral dissertations. We have also used material from the Master's theses of Atanu Phukan, Srikrishna Sahu, A. A. Kakade, Kaladhar Semwal, Rakesh Ranjan, and Vikas Kumar.

Alok Prasad, Ankur Sharma, Yogendra Rathi, B. R. Vinoth, Prateek Khanna, Abhinav Parashar, Yogesh Nimdeo, and Veena Singh helped us with figures and we thank them for their time.

Financial support from funding agencies allowed us to make the apparatus and configure the measurement systems reported in this book. We gratefully acknowledge support received from the Department of Science and Technology, New Delhi, Board of Research in Nuclear Sciences, Mumbai, and the Ministry of Human Resource Development, New Delhi.

We thank Professor Frank Kulacki, Series Editor, *Thermal Science Briefs* for providing us this opportunity and continuous encouragement.

We are grateful to our Institute for the ambience it provides and our families for strong support.

Contents

Chapter 1
Optical Methods: An Overview

Keywords Laser · Interference · Coherence · Fringe formation · Refraction · Lorentz–Lorenz formula

1.1 Introduction

This chapter introduces the basic ideas involved in optical measurement techniques for characterizing heat and mass transfer processes in fluids. The fluid medium is taken to be transparent to the passage of light. The light intensity distribution and contrast generated by the measurement depend on changes in refractive index in the field-of-interest. Applicable optical techniques such as interferometry, schlieren, and shadowgraph are interrelated. Thus, the present chapter aims at setting up principles of interferometry. Laser is an important light source and is an indispensable part of an optical measurement. A short introduction to lasers and their classification is also provided. Interferometry is limited by beam bending errors that in turn form the foundation of schlieren and shadowgraph techniques, discussed in the following chapters.

1.2 Background

Optical methods of measurement are known to have specific advantages in terms of spanning a field-of-view and being inertia-free. Although in use for over half a century, optical methods have seen a resurgence over the past decade. The main factors responsible are the twin developments in the availability of cost-effective lasers along with high performance computers. Laser measurements in thermal sciences have been facilitated additionally by the fact that fluid media are transparent and heat

P. K. Panigrahi and K. Muralidhar, *Schlieren and Shadowgraph Methods in Heat and Mass Transfer*, SpringerBriefs in Thermal Engineering and Applied Science, DOI: 10.1007/978-1-4614-4535-7_1, © The Author(s) 2012

transfer applications in fluids are abundant [1–7]. Whole-field laser measurements of flow and heat transfer in fluids can be carried out with a variety of configurations: shadowgraph, schlieren, interferometry, speckle, and PIV, to name a few. In the present monograph, temperature and concentration field measurement in fluids mainly by schlieren has been addressed.

The ability to record a time sequence of optical images in a computer using CCD cameras and frame grabbers has permitted data analysis. In addition, it is possible to enhance image quality and perform operations such as edge detection, fringe thinning, and contrast improvement by manipulating the light intensity values representing the image.

When combined with tomography, laser measurements can be extended to map 3D temperature and concentration fields. Here, the optical images are viewed as projection data of the thermal and species concentration fields. The 3D field is then reconstructed by suitable algorithms. In principle, tomography can be applied to a set of projection data recorded by shadowgraph, schlieren, interferometry, or any of the other configurations [6–8]. In the present monograph, tomography as applied to schlieren data is described. These comments carry over to interferometry and shadowgraph methods as well.

In *transparent* media, the interaction of light with the material is via the refractive index n defined as

$$n = \frac{c}{c_0}$$

Here, c_0 is the speed of light in vacuum and c is the speed of light in the transparent medium. It can be shown that the refractive index satisfies the inequality $n \geq 1$. The lower limit $n = 1$ is reached in vacuum. The utility of refractive index in measurements arises from the fact that, for isotropic transparent media, it is a unique function of material density. Since density, in turn, will depend on temperature and species concentration, inhomogeneities in the refractive index field carry information related to heat and mass transfer processes.

It is important to comment on the physical phenomenon called *scattering*, namely, the interaction of light with matter. When a beam of light (wavelength λ) falls on a particle of size d_p, its characteristic dimension, the scattered energy will show changes with respect to intensity, directionality, wavelength, phase, polarization, and other properties of the wave. The property that shows the most pronounced change depends on the ratio of the wavelength and the particle dimension. Broadly speaking, we have the following limits:

1. Ray optics

$$\frac{\lambda}{d_p} \ll 1$$

2. Wave optics

$$\frac{\lambda}{d_p} \approx 1$$

3. Quantum optics

$$\frac{\lambda}{d_p} \gg 1$$

Measurement techniques such as laser Doppler velocimetry, particle image velocimetry, laser-induced fluorescence, and Rayleigh and Raman scattering depend on the interaction of light with matter, but do not fall within the purview of this monograph.

The discussions in the present text deal with transparent media and images originate from the distribution of refractive index in the field-of-interest. Optical configurations such as interferometry, schlieren, and shadowgraph are interrelated and exploit changes in refractive index. Among the three, the focus here is on schlieren and, to some extent, shadowgraph imaging of transport phenomena.

1.3 Optical Methods

Refractive index techniques exploit the wave nature of electromagnetic radiation for measuring temperature and concentration distributions. Optical imaging works with the visible range of the electromagnetic spectrum (wavelength $\lambda = 400\text{--}700\,\text{nm}$) and optical images are visible to the naked eye.

Optical effects are associated primarily with the electric field rather than magnetic. Hence, for analysis, the propagation of light originating from a light source can be taken as

$$E = \sum_j A_j \sin\left(\frac{2\pi}{\lambda_j}(ct - x) + \phi_j\right) \tag{1.1}$$

Here, E is the strength of the electric field, A_j is its amplitude corresponding to wavelength λ_j, ϕ_j is the phase associated with the jth harmonic, c is the speed of light and t, and x are time and position coordinate respectively. For a white light source, electronic transitions are random in time. As a result, the phases of the respective harmonics are also random quantities. For a monochromatic source, only one wavelength is significant, phase may be set to zero, and we get

$$E = A \sin\left(\frac{2\pi}{\lambda}(ct - x)\right) \tag{1.2}$$

The electric field E is, in general, a vector but in measurements one works with a single light beam or beams that are nearly parallel. Hence, it is sufficient to examine the magnitude of E. Two monochromatic wave fronts arising from a single light source and having a phase difference ϕ are represented by the equations

$$E_1 = A \sin\left(\frac{2\pi}{\lambda}(ct - x)\right)$$

$$E_2 = A \sin\left(\frac{2\pi}{\lambda}(ct - x) + \phi\right) \qquad (1.3)$$

Interferometric measurements are based on information contained in these equations. The phase difference ϕ is equivalent to a path difference δ where

$$\delta = \frac{\lambda}{2\pi}\phi$$

Here, δ is called as the optical path length, to be contrasted with the geometric path length represented by the x-coordinate. Geometric path length forms the basis of distance measurement while the phase difference forms the basis of distance, speed, density, and temperature measurements. These measurements are, however, possible only if the phase difference is stable and independent of time. This condition on phase requires the light source to be *coherent*. The laser is a high quality monochromatic coherent light source and is appropriate for optical instrumentation.

1.4 Light Sources

Conventional light sources emit radiation by a series of successive spatially distributed random phenomena. These involve atomic excitation followed by emission as electrons in the valence band jump to lower energy levels. The downward transition follows predetermined rules that finally govern the wavelength of the emitted photon. The energy emitted e and wavelength λ are related as $e = hc/\lambda$, where h is Planck's constant. The average life of a radiating atom is 1.6×10^{-8}s and the average length of a single train of waves is $CL = cT = 4.8$ m. Coherence length of a tungsten filament is at best a fraction of a mm. In contrast, a laser produces a beam whose coherence length can be a fraction of a meter. Further, a laser beam is thin that effectively originates from a point source.

1.4.1 Introduction to Lasers

We describe briefly the principles involved in the operation of a laser. Laser, an acronym for *light amplification by stimulated emission of radiation*, employs the following ideas.

    a. *Metastable States*. Normally, valence electrons of an atom can be excited to a higher energy level from which they return to the ground state by emitting a photon. However, for certain materials there exist energy levels beyond the ground state from which the return of electrons to the ground state is considerably delayed. This return can, however, occur in the event of a collision between an electron and a photon. The average life of a normally excited electron is 10^{-8} s; in the metastable state it can be as high as 10^{-2} s.

b. *Optical Pumping*. It is possible to raise electrons to metastable states by light absorption. This process is called optical pumping.

c. *Fluorescence*. Emission of light when an electron jumps from a metastable state to the ground state is called *fluorescence*. In light sources, the gas pressure is kept very low to minimize the possibility of collision between particles, increase the particle life in the metastable state, and minimize the production of thermal energy. Hence, the resulting electronic transitions mainly classify as fluorescence.

d. *Population Inversion*. When the number of electrons in the metastable state exceeds that in the ground state, one obtains population inversion.

e. *Resonance*. In the absence of collisions, electrons undergoing transition from the metastable state will emit radiation at the frequency that is equal to the absorption frequency, a phenomenon called resonance.

f. *Stimulated Emission*. A photon released during emission from an electron in the metastable state can stimulate another high energy electron to release a photon of identical frequency, direction, polarization, phase, and speed. The two photons are now completely identical and radiation is both temporally and spatially coherent. Stimulated emission and stimulated absorption in the ground state are equally probable. For a net emission, i.e., to construct a light source it is important to have a population inversion.

g. *Cavity Oscillation*. Starting with a gas with metastable states and population inversion, a light source can be constructed. However, the effectiveness of stimulation is increased by confining the gas between parallel mirrors. Here, reflections lead to increased stimulation and a large number of photons will surge back and forth in the cavity formed by the mirrors. Further, the wavelength λ_L for which the cavity length $L = N\lambda_L$ with N being an integer will produce a standing wave pattern whose wavelength is λ_L. Other wavelengths will be dissipated as thermal energy. A small round opening in the reflection coating of one of the mirrors will then enable the passage of monochromatic (plane polarized) radiation of wavelength λ_L that can be used in measurements.

h. *Helium–Neon Laser*. A He–Ne laser is a popular light source in optical instrumentation. It has a continuous wave output typically in the range of 0.5–75 mW and a wavelength of 632.8 nm. It is sturdy in construction, economical and stable in operation. A sketch of this laser is given in Fig. 1.1. Here, 1 and 2 are fully and partly silvered mirrors, respectively and form a cavity. V is a high voltage DC source used to excite the helium atoms. Gas in the cavity is at a low pressure; the partial pressures of He and Ne are 1 and 0.1 mm of mercury column (1/300 and 1/3000 of the standard atmospheric pressure). Lasing action is related to stimulated emission of the neon atoms. Helium atoms are excited to metastable states by the externally applied voltage. These can return to the ground state by collision with unexcited neon atoms. In the process, the upper most energy levels of neon get populated. As neon atoms return to lower energy levels, the photons released stimulate additional radiation from excited neon atoms and the entire process is intensified by the

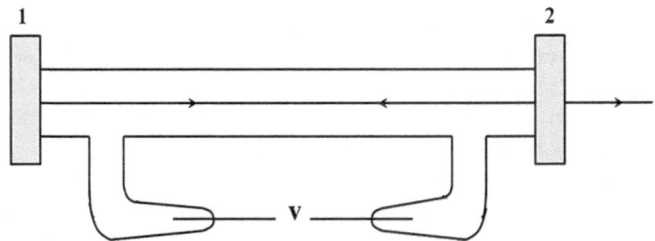

Fig. 1.1 Construction of a helium–neon laser. Mirrors 1 and 2 form the optical cavity. Voltage V from a high voltage DC source is used to excite helium atoms to higher energy levels

reflecting cavity walls. The wavelength of 632.8 nm (an orange–red color) is chosen over other wavelengths by proper choice of the cavity length.

1.4.2 Characteristics of Lasers

In the context of measurements, lasers fall in the category of light sources with certain helpful properties. Unlike a material probe (say, a thermocouple), lasers are often called *photon probes*. The measurement of a flow property may, however, rely on the wave-like nature of light. Although light propagates as a wave, generation of light from a material is a quantum-mechanical phenomenon. In fact, the subject is one of emission of electromagnetic radiation, with light being EM radiation in the visible range (400–700 nm, wavelength). In a conventional light source, for example a tungsten filament lamp, the metallic wire is electrically heated to a high enough temperature. Under a thermal stimulus, electrons undergo transition to higher energy levels. As they return to the ground state, they emit photons. Transitions occurring closer to the surface of the material result in a net emission of radiation to the environment. If the filament temperature is high enough, photons would be energetic and radiation can fall in the visible range. Unfortunately, such a light source has limited utility in measurements.

The tungsten filament lamp is often called a *conventional light source*, to contrast its behavior with a laser. Characteristically, the filament emission is random in time, spatially distributed, and photon energies are distributed over several wavelengths, thus giving rise to polychromatic radiation. The randomness in time ensures that the phases of the wave packets leaving the filament are practically uncorrelated. Further, the emission is in all directions and intensity diminishes with distance. In contrast, a laser output is

1. monochromatic,
2. intense,
3. directional, and
4. coherent.

Table 1.1 Various types of lasers and their overall specification (after [2])

Medium	Phase	Mode	Wavelength λ, nm	Power	Energy per pulse	Coherence length (cm)
He–Ne	Gas	Continuous	632.8 (orange–red)	0.1–75 mW	-	20
Argon-Ion	Gas	Continuous	488 (blue) 514 (green)	0.1–10 W		5 2000 (with etalon)
Krypton	Solid	Continuous	47–676	0.1–0.9 W		5–18
Ruby	Solid	Pulsed or continuous	694 nm	0.1–1 W	500–2000 mJ	50–500 (with etalon)
Nd:YAG	Solid	Pulsed	1064		0.1–100 J	1
CO_2	Gas	Pulsed or continuous	1062	10 kW	2000 mJ	Small

An *etalon* is an optical interferometer included with the laser and helps improve coherence of the light output

Fig. 1.2 Intensity I as a function of phase difference ϕ between two interfering light beams

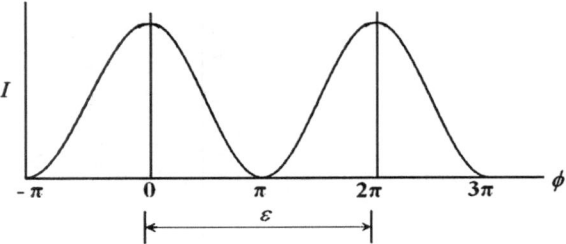

A summary of various commercially available lasers is given in Table 1.1. A helium–neon laser is most popular in measurements. For measurements in liquids and when multiple lines (wavelengths) are required, the argon-ion laser is preferred. This is because of its superior coherence at higher power outputs. A CO_2 laser is unsuitable in the measurement context but is preferred in the manufacturing industry where operations such as drilling and cutting are common.

A summary of optical measurement techniques employed in various applications is provided in Table 1.2. Examples where white light is used are included for completeness.

1.5 Interference Phenomenon

Consider the superposition of two nearly parallel electromagnetic waves that originate from a given monochromatic light source; the waves have a phase difference ϕ. Their amplitudes are taken to be equal. Superposition of these wavefronts leads to the form [3]

$$
\begin{aligned}
E_1 + E_2 &= A\left[\sin\left(\frac{2\pi}{\lambda}(ct-x)\right) + \sin\left(\frac{2\pi}{\lambda}(ct-x)+\phi\right)\right] \\
&= 2A\cos\frac{\phi}{2}\sin\left(\frac{2\pi}{\lambda}(ct-x)-\frac{\phi}{2}\right)
\end{aligned}
\tag{1.4}
$$

The intensity I of the combined beam is $4A^2\cos^2\phi/2$ and is plotted in Fig. 1.2. To the human eye, intensities below a certain threshold would be seen as dark while intensities above would be bright. Light sensors can, of course, detect small changes in intensity. To an observer, the intensity distribution of Fig. 1.2 is a sequence of dark and bright patches, called *fringes*.

With reference to Fig. 1.2, the superposition of two light beams with uniform intensity but a phase difference results in an interference pattern consisting of alternately dark and bright regions, namely, the fringes. The spacing between two lines corresponding to the highest intensity is called a fringe shift and is marked ϵ in the figure. This fringe shift is also obtained as the spacing between adjacent lines

Table 1.2 Overview of optical methods suitable for measurement of velocity, temperature and concentration [5]

Method (physical mechanism)	Incident light	Attribute measured	Quantities detected	Real-time	Spatial extent
Interferometry (refractive index)	Laser beam	Intensity	T and C	Yes	2D (integrated)
Schlieren and shadowgraph (refractive index)	Laser beam	Intensity field	T and C	Yes	2D (integrated)
Rainbow schlieren (refractive index)	White light	Hue	T and C	Yes	2D (integrated)
LDV (elastic scattering)	2 laser beams per dimension	Doppler shift	$\mathbf{u}$	Yes	Point
PIV (Mie scattering)	Laser sheet	Intensity	$\mathbf{u}$	Yes	2D and 3D
LCT (Mie scattering)	White light	Hue	T and wall shear	Yes	2D
LIF (fluorescence)	Laser Sheet	Intensity	T and C	Yes	2D
CARS (inelastic scattering)	2 laser beams	Intensity spectrum	T and C	No	Point

LDV is laser Doppler velocimetry; PIV is particle image velocimetry; LCT is liquid crystal thermography; CARS is coherent anti-Stokes Raman spectroscopy; and LIF is laser-induced fluorescence. Temperature and concentration are represented as T and C respectively while $\mathbf{u}$ is the velocity vector

Fig. 1.3 Definition of temporal coherence between points 3 and 4 and spatial coherence between points 1 and 2

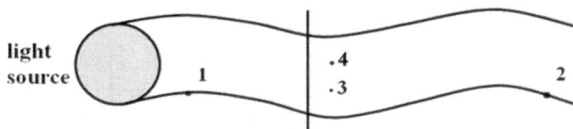

of minimum intensity. Since intensity varies as $\cos^2 \phi/2$, the corresponding phase difference for a complete fringe shift is 2π. The equivalent optical path difference is λ. The interference fringes are ordered, i.e., the nth fringe counted from the left will represent a phase difference of $2n\pi$ with respect to the reference wave. In distance measurements, an integer number of fringe shifts is used and the resolution in distance measurement is λ.

In a given problem, individual portions of the light beam traverse distinct optical paths and the phase difference is a spatially distributed variable. In an interferometric measurement, this beam is combined with one that has a constant phase. In regions where the phase difference between the two is $(2n - 1)\pi, n = 1, 2...$, the intensity of the combined beam is zero and we get *destructive* interference. When the phase difference is $2n\pi$ we get *constructive* interference. The corresponding path differences are $(2n - 1)\lambda/2$ and $n\lambda$, respectively. The phase field in the form of a fringe pattern can be recorded, say by a camera, to extract information about the primary variables of the problem being studied. In applications, quantitative measurements are possible if lines of *constant phase* exist so that fringes form, and a fringe shift can be identified from one fringe to another.

From Eq. 1.4, it follows that an interference pattern is stable only if the following conditions hold:

1. Each wave train has traversed the same geometric distance $x = L$.
2. Phase at a point and the phase difference between two points are constant with respect to time.

Point 1 requires that the light source be a point. Point 2 requires that the source emit a wave continuously and endlessly for all time. These requirements are met to a great extent by a laser but not a conventional light source.

Consider the propagation of light in a homogeneous medium as sketched in Fig. 1.3. For the interference pattern to be stable in time, we require $\phi_3 = \phi_4$ and $\phi_2 - \phi_1$ to be independent of time. These conditions are ones of spatial and temporal coherence, respectively. Conventional light sources such as a tungsten filament lamp emit sporadically and phase here is a random variable. Phase quality is a special feature of lasers.

Instruments that utilize interference phenomenon for measurements are called *interferometers*. A stable interference pattern in which the fringes are ordered is a prerequisite in all interferometry measurements. However, no light source is perfectly coherent and the quality of a light source is judged by its coherent length CL, i.e., the maximum distance between points 1 and 2 in Fig. 1.3, over which $\phi_2 - \phi_1$ is time independent. Helium–neon lasers have a coherence length of 100–200 mm and

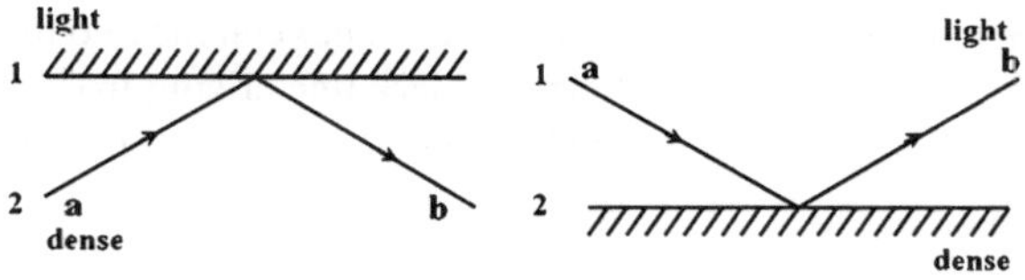

Fig. 1.4 Phase differences arising from reflection at material interfaces. Medium 2 is denser than 1

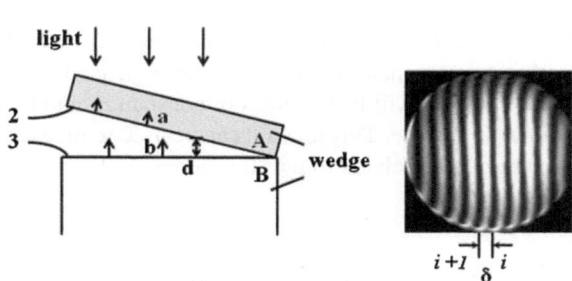

Fig. 1.5 Wedge fringes (*right*) created by an air gap of varying thickness (*left*). Ray *a* is reflected from the inner partially silvered surface of the block. Ray *b* reflects from the fully silvered lower surface. Distance *d* is the gap between the block and lower surface

are considered suitable for interferometry. An argon-ion laser with etalon can have a higher coherence length, higher power, and can also generate two wavelengths in the visible range.

The following results are important for the calculation of phase difference between two light beams. Consider reflection of a ray of light as it approaches a denser material from a lighter material (Fig. 1.4, right). With ρ as density and n, the refractive index, we have, $\rho_1 < \rho_2$, and $n_1 < n_2$. Then, it can be shown that $\phi_b - \phi_a = \pi$, where ϕ is the phase angle.

Consider reflection of light as it approaches a lighter material from the denser side (Fig. 1.4, left). We have, $\rho_1 < \rho_2$ and $n_1 < n_2$. Then $\phi_a = \phi_b$, i.e., the reflected ray has the same phase as the incident ray.

As an example, consider the formation of wedge fringes in the arrangement shown in Fig. 1.5. A transparent block A rests on another block B and the spacing d between them varies with position. Surface 2 is partly silvered, surface 3 is fully silvered and these surfaces are exposed to light at normal incidence. The interference pattern arising from the superposition of rays a (leaving A after reflection) and b (leaving B after reflection) is shown in Fig. 1.5. If the incident light has a phase ϕ then $\phi_a = \phi$ and

$$\phi_b = \phi + \pi + 4\frac{\pi d}{\lambda}$$

Here, the phase difference arising from the block A can be ignored because its contribution is common to rays a and b. Hence,

$$\phi_a - \phi_b = 4\frac{\pi d}{\lambda} + \pi$$

When $d = 0$, the phase difference between the two light beams is π and the first line of the interference pattern is a dark line arising from destructive interference. As we go from the ith to $(i+1)$th fringe, d changes from d_i to $d_i + 1$, $\phi_a - \phi_b = 2\pi$ and so

$$d_{i+1} = d_i + \frac{\lambda}{4}$$

The factor 1/4 is related to the fact that ray b traverses the distance d twice. The above method permits measurement of d starting from the first dark fringe at the right where $d = 0$. Fringes will follow lines of constant d, and hence represent contour lines on an uneven surface. This method can be used to measure d or the flatness of a surface, the latter being related to the straightness of the wedge fringes.

1.5.1 Temperature Measurement

In measurements related to heat and mass transfer, changes in the phase of a light beam originate from the variation in the refractive index of the medium itself. The refractive index of a transparent material is defined as

$$n = \frac{c(\text{vacuum})}{c(\text{material})}$$

where c is the speed of light. The reduction in the speed of light in matter can also be viewed as an increase in the optical path length to be covered by the electromagnetic waves, and hence a source of phase difference. The density and refractive index of a transparent material are uniquely related, and to a leading order, they are connected by the *Lorentz–Lorenz* formula [1]

$$\frac{n^2 - 1}{\rho(n^2 + 2)} = \text{constant} \qquad (1.5)$$

In gases $n \approx 1$ and the relationship simplifies to

$$\frac{n - 1}{\rho} = \text{constant}$$

Hence, in gases, the derivative

$$\frac{dn}{d\rho} = \text{constant}$$

In liquids, the derivative is nearly constant if the bulk changes in density are small.

For moderate changes in temperature (say, up to, around $10\,^\circ\text{C}$ in air), density and temperature T are linearly related as

Fig. 1.6 Coordinate system within the fluid medium where $x-y$ is the cross-sectional plane and z, the direction of propagation of light. The distance L is the spatial extent of the test section in the viewing direction

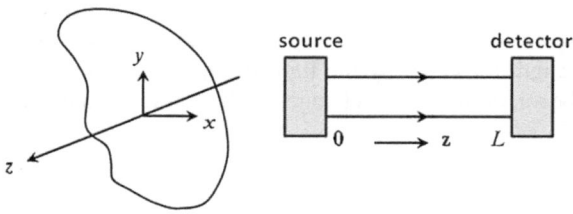

$$\rho = \rho_0(1 - \beta(T - T_0)), \quad \beta > 0$$

It follows that

$$\frac{dn}{dT} = \text{constant}$$

and changes in temperature will simultaneously manifest as changes in the refractive index. This result carries over to mass transfer problems as well, where density changes occur from a solutal concentration field.

Remarks:

1. Refractive index depends on the wavelength of light and accordingly, the constant in Eq. 1.5 and the equations that follow are functions of wavelength. The data quoted in the present monograph refer to helium–neon laser as a light source.
2. The dependence of refractive index on wavelength is often given by the Cauchy's formula [9]

$$n(\lambda) = A + \frac{B}{\lambda^2} + \frac{C}{\lambda^4} + \cdots$$

The dependence is stronger in liquids and solids when compared to gases. The first two terms may be adequate in most applications.

3. The Lorentz–Lorenz formula is a special case of a more general dependence function of refractive index on density. The general form is written in the form of a virial expansion in the dependent variables (ρ, T, C) as follows:

$$\frac{n^2 - 1}{\rho(n^2 + 2)} = a_0 + a_1\rho + a_2 T + a_3 C$$

The individual dependence on temperature and concentration (apart from density) is usually small for most media. Hence, it is sufficient to generalize Eq. 1.5 as

$$\frac{n^2 - 1}{\rho(n^2 + 2)} = a_0 + a_1\rho$$

The correction has negligible effect in gases. In liquids, the derivative $dn/d\rho$ may be altered by 5–10 % when the second term is included in calculations.

Consider a beam of light moving through a gaseous medium of varying temperature and total length L (Fig. 1.6). The distance L is also equal to the geometric path length traversed by the light beam. The optical path length traversed by the light beam, corrected for changes in the refractive index is

$$PL = \int_0^L n\,dz$$

The integral is greater than L since $n > 1$ (except in absolute vacuum, where the integral is simply L). Suppose light beam 1 propagates through a region of variable density and hence refractive index, n while beam 2, through a region of constant density (and refractive index). The difference in path lengths between 1 and 2 can be calculated using the first-order Taylor series approximation as

$$\Delta PL = PL_1 - PL_2 = \int_0^L (n_1 - n_2)\,dz$$

$$= \frac{dn}{d\rho} \int_0^L (\rho_1 - \rho_2)\,dz$$

$$= \frac{dn}{dT} \int_0^L (T_1 - T_2)\,dz \tag{1.6}$$

In problems where T_1 is a 2D temperature field $T_1(x, y)$, the integral simplifies to

$$\Delta PL = (T_1 - T_2)\frac{dn}{dT}L \tag{1.7}$$

Since a path length difference of λ will generate one fringe shift, the temperature difference ΔT_ϵ required for this purpose is

$$\Delta T_\epsilon = \lambda \Big/ \left(L\frac{dn}{dT}\right) \tag{1.8}$$

In air, $dn/dT = -0.927 \times 10^{-6}/°C$; in water, $dn/dT = -0.88 \times 10^{-4}/°C$. Since density (mostly) decreases with increasing temperature at constant pressure, both derivatives are negative under normal conditions.

For a He–Ne laser $\lambda = 632.8$ nm. Hence, $L \times \Delta T_\epsilon$ per fringe shift for a He–Ne laser is $0.682\,°C$-m in air and $0.0072\,°C$-m in water. Note that, the value of ΔT_ϵ itself decreases with increasing geometric path length L. Hence, the sensitivity of measurements can be adjusted by designing apparatus of suitable dimensions in the direction of propagation of light. In high temperature application L is made small while in problems involving small temperature differences L can be large. The largest value of L is, however, limited by refraction and higher order errors (Sect. 1.4.5).

The above discussion shows that each fringe of an interferogram is a line of constant phase, constant refractive index, constant density, and hence constant temperature, namely an isotherm. This interpretation is useful in qualitative interpretation of interference patterns.

Since higher order terms of the Taylor series expansion are neglected, Eqs. 1.6– 1.8 hold for small changes in density. The correctness of the approximation calls for an independent validation from experiments. The expression for optical path length further assumes that light rays travel is straight lines. In a more general setting, light rays would refract and the path of light propagation would be a curve $s = s(x, y, z)$. The optical path length is then evaluated along this curve as

$$PL = \int n(x, y, z))\mathrm{d}s \tag{1.9}$$

1.5.2 Dual Wavelength Interferometry

In experiments involving changes in temperature as well as concentration, a *dual wavelength* laser is used to record two sets of interferograms [4]. Dual wavelength interferometry can separate the influence of two competing parameters, such as temperature and concentration, provided the sensitivities $\partial n/\partial T$ and $\partial n/\partial C$ are sufficiently different at the two wavelengths. This problem is similar to solving two equations for two unknown variables. For processes influenced by changes in temperature and concentration, the change in refractive index is written as

$$\Delta n = \frac{\partial n}{\partial T}\Delta T + \frac{\partial n}{\partial C}\Delta C \tag{1.10}$$

Measurements are carried out from one fringe to another. The change in refractive index at a fringe of order N in an apparatus of length L in the viewing direction is obtained as

$$\Delta n = \frac{N\lambda}{L}$$

Hence, for each wavelength λ_1 and λ_2 with corresponding refractive indices n_1 and n_2, we get

$$\frac{N\lambda_1}{L} = \frac{\partial n_1}{\partial T}\Delta T + \frac{\partial n_1}{\partial C}\Delta C$$

$$\frac{N\lambda_2}{L} = \frac{\partial n_2}{\partial T}\Delta T + \frac{\partial n_2}{\partial C}\Delta C$$

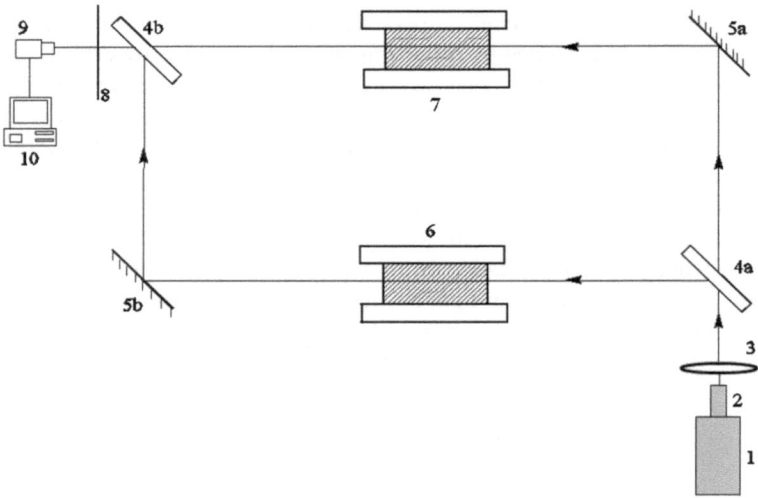

Fig. 1.7 Schematic drawing of a two beam Mach–Zehnder interferometer. *1* Laser; *2* Spatial filter; *3* Convex (or plano convex) lens; *4(a)*, *4(b)* Beam splitters; *5(a)*, *5(b)* Mirrors; *6* Test cell; *7* Compensation chamber; *8* Screen; *9* CCD Camera; *10* Computer

These equations can be simultaneously solved for ΔT and ΔC, the changes in temperature and concentration in the physical domain.

1.5.3 Mach–Zehnder Interferometer

It is a popular instrument used in experimental studies of heat and mass transfer in fluids. The experimental apparatus is located in the path of test beam 2 shown in Fig. 1.7. Quantitative experiments are possible when the temperature (or concentration) field is 2D in the cross-sectional plane and uniform parallel to the beam direction. Otherwise, the test beam integrates the temperature field as it traverses the test cell and only qualitative information can be obtained. The reference beam 1 passes through a region identical to the test cell except that the fluid here is at a uniform temperature. For experiments in air, the reference beam may pass through the normal ambient and no special treatment is adopted. In liquids, a *compensation chamber* containing the liquid constant temperature (or concentration) is required so that the optical path difference is entirely due to changes in temperature (or concentration) alone. The spatial filter expands the laser beam that is subsequently collimated using a convex lens. The spatial filter along with the lens constitutes the collimating arrangement of the interferometer.

The initial (geometric) path lengths of beams 1 and 2 are nearly equal except for possible angular misalignment of the mirrors and the beam splitters. As these

Fig. 1.8 Fringes (*right*)
formed due to misalignment
of optics (*left*) in a Mach–
Zehnder interferometer. *Solid
and dashed lines* on the
left indicate aligned and
misaligned positions of the
optical elements

are turned, the interferometer approaches a state of complete alignment. The initial
fringe pattern corresponds to wedge fringes whose spacing increases as the optical
alignment improves. The best initial setting corresponds to two fringes widely apart,
containing the full field-of-view. This is called the *infinite-fringe* setting. Every point
in reference beam and the test beam has the same path length measured from the
laser to the screen. A thermal disturbance placed in the path of the test beam will
produce a fringe pattern where each fringe is an isotherm.

The Mach–Zehnder interferometer is not always used with the two interfering
wavefronts parallel to each other, as in the infinite-fringe setting. There is a second mode in which the two interfering wavefronts have a small angle θ between
them, introduced deliberately during alignment. Upon interference they produce an
image consisting of bright and dark fringes, representing the loci of constructive and
destructive interference, respectively (Fig. 1.8). These parallel and equally spaced
fringes are referred to as *wedge* fringes. The terminology follows the discussion on
fringe formation in a variable air gap (Sect. 1.4). The spacing d_w between the wedge
fringes is a function of the tilt angle θ and the wavelength λ of the laser used and is
given by

$$d_w = \frac{\lambda/2}{\sin\theta/2} \approx \frac{\lambda}{\theta}$$

When a thermal disturbance is introduced in the path of the test beam, fringes deform
to an extent that depends on the change in temperature at that location. Hence, imaging
using wedge fringes yields temperature profiles prevailing in the physical domain.

1.5.4 Fringe Analysis

Interferograms recorded above a candle flame in the infinite and wedge fringes set-
tings are shown in Fig. 1.9a. The first and the third image from the left are for
undisturbed conditions, while the second and the fourth are in the plume of a candle
flame.

In the absence of a thermal disturbance, the infinite-fringe setting portrays a clear
bright field, except for stray fringes that may arise from imperfections in the optical
elements. Newer fringes are created when the medium is thermally perturbed. Each
pair of fringes represents a temperature shift of ΔT_ϵ. The spacing between fringes

will depend on the temperature gradient prevailing at that location. Let T_w be the wall temperature and T_1, the temperature of the fringe next to it. Let δ represent the distance between the fringe and the wall; in most applications, δ will be spatially distributed. Near a wall, the heat flux exchanged by the surface with the fluid of thermal conductivity k_f is calculated as

$$q_w = -k_f \frac{\partial T}{\partial y} \approx -k_f \frac{T_w - T_1}{\delta}$$

In the wedge fringe setting, the fringes are initially straight. When the wall is heated, fringes displace by an amount d that depends on the change in temperature at that point. Let D be the maximum fringe displacement, at the wall, for the present discussion. Temperature profile is obtained as

$$\frac{T - T_{\text{ambient}}}{T_{\text{wall}} - T_{\text{ambient}}} = \frac{d}{D}$$

Wall heat flux is calculated as

$$q_w = -k_f \left. \frac{\partial T}{\partial y} \right|_{y=0}$$

Fringe patterns during buoyancy-driven convection in an eccentric annulus are shown in Fig. 1.9b.

1.5.5 Refraction Errors

The derivation of the expression for temperature difference between successive fringes (Eqs. 1.6–1.8) assumes that the passage of the light beam is along straight lines and will be modified in the presence of a strongly refracting field. Refraction leads to beam bending and increases the optical path length to the test beam. In addition, the path length determination will require integration of refractive index along the true path of the test beam. Following [1], the extent to which Eqs. 1.6–1.8 are altered by refraction can be estimated under simplified conditions as follows.

Consider the path of the light ray AB through a test cell (Fig. 1.10) when it is affected by the refraction effects. Refractive index is taken to vary mainly in the y-direction while the propagation of light is along the z-direction. Variations in the x-direction are neglected. Let α be the bending angle at a location P of the test cell. The optical path length from A to B is given by

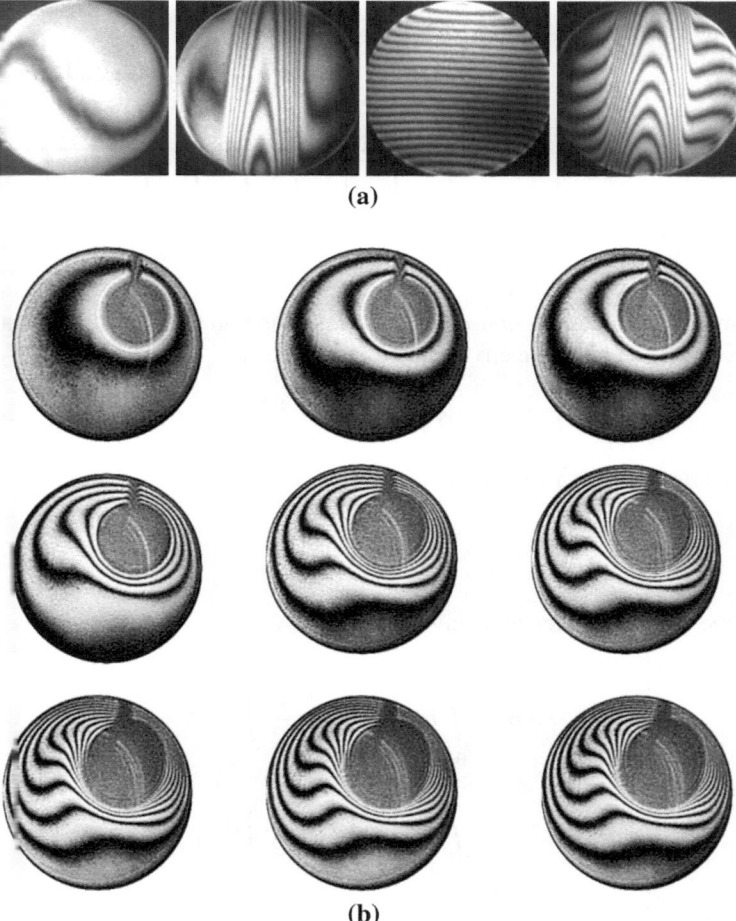

Fig. 1.9 *a* Fringes recorded in a plume above a candle flame in an infinite-fringe setting (*left*) and the wedge fringe setting (*right*). *b* Time sequence of interferometric fringes formed during buoyancy-driven convection in air within an eccentric annulus, with the inner cylinder heated and the outer cylinder cooled

$$AB = \int_0^L n(y, z) \mathrm{d}s$$
$$= \int_0^L n(y, z) \frac{\mathrm{d}z}{\cos \alpha}$$

Fig. 1.10 Path of a ray of
light in a region with refractive index variation in the
transverse (y-) direction

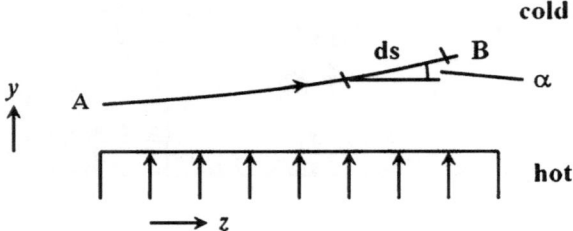

To examine the immediate consequence of refraction on the optical path length, we
assume α to be small while $\cos\alpha$ can be expressed as

$$\cos\alpha = (1 - \alpha^2)^{1/2}$$

Using the first two terms of the binomial expansion

$$\cos\alpha \approx 1 - \frac{\alpha^2}{2}$$

Hence the optical path length is given by

$$AB = \int_0^L n(y, z)\left(1 - \frac{\alpha^2}{2}\right)^{-1} dz$$

$$= \int_0^L n(y, z)\left(1 + \frac{\alpha^2}{2}\right) dz \tag{1.11}$$

Angle $\alpha(z)$ at any location z can be calculated using Snell's law as:

$$\alpha(z) = \int_0^z \frac{1}{n(y, \tilde{z})} \frac{\partial n(y, \tilde{z})}{\partial y} d\tilde{z}$$

Also see Eq. 2.5. To a first approximation

$$\alpha(z) = \frac{1}{n(y, z)} \frac{\partial n(y, z)}{\partial y} \times z$$

From Eq. 1.11, the optical path length from A to B is

$$AB = \int_0^L n(y,z)\left(1 + \frac{1}{2}\frac{1}{n^2}\left(\frac{\partial n}{\partial y}\right)^2 z^2\right) dz$$

$$= \bar{n}(y)L + \frac{1}{6\bar{n}(y)}\left(\frac{\partial \bar{n}}{\partial y}\right)^2 L^3$$

Here $\bar{n}(y)$ is the average of $n(y,z)$ over the length L, determined along the z-coordinate. Similarly the expressions $1/\overline{6n}$ and $\partial\bar{n}/\partial y$ represent average line integrals over the length L. The optical path length of the reference beam is simply

$$\text{Reference path length} = \int_0^L n_0 dz = n_0 L \tag{1.12}$$

The difference in the optical path lengths of the test and the reference beams in the presence of refraction effects is

$$\Delta PL = \bar{n}(y)L + \frac{1}{6\bar{n}(y)}\left(\frac{\partial \bar{n}}{\partial y}\right)^2 L^3 - n_0 L$$

$$= (\bar{n}(y) - n_0)L + \frac{1}{6\bar{n}(y)}\left(\frac{\partial \bar{n}}{\partial y}\right)^2 L^3$$

$$= (\bar{T_1}(y) - T_0)L\frac{dn}{dT} + \frac{1}{6\bar{n}(y)}\left(\frac{\partial \bar{n}}{\partial y}\right)^2 L^3$$

where $\bar{T_1}(y)$ represents the line integral of the temperature field along the direction of the ray at a given point on the fringe, divided by L. The corresponding ray over the next fringe traverses an additional distance of λ. Hence, one can write

$$\Delta PL + \lambda = (\bar{T_2}(y) - T_0)L\frac{dn}{dT} + \frac{1}{6\bar{n}(y)}\left(\frac{\partial \bar{n}}{\partial y}\right)^2 L^3$$

where $\bar{T_2}(y)$ represents the average line integral of the temperature field along the direction of the ray at a point on the next fringe. The temperature difference between successive fringes can be computed from

$$\lambda = (\bar{T_2}(y) - \bar{T_1}(y))L\frac{dn}{dT} + \frac{1}{6\bar{n}(y)}\left(\frac{dn}{dT}\right)^2\left(\left(\frac{\partial \bar{T}}{\partial y}\Big|_2\right)^2 - \left(\frac{\partial \bar{T}}{\partial y}\Big|_1\right)^2\right)L^3 \tag{1.13}$$

The temperature drop per fringe shift is

$$\Delta T_\epsilon = \frac{1}{L\,dn/dT}\left(\lambda - \frac{1}{6n(y,z)}\left(\frac{dn}{dT}\right)^2\left(\left(\frac{\partial\overline{T}}{\partial y}\bigg|_2\right)^2 - \left(\frac{\partial\overline{T}}{\partial y}\bigg|_1\right)^2\right)L^3\right)$$

(1.14)

Since the gradient in the temperature field is not known before the calculation of the fringe temperature the factor containing the temperature derivatives must be calculated from a guessed temperature field. Thus, the final calculation of ΔT_ϵ relies on a series of iterative steps with improved estimates of the temperature gradients.

The formulation of temperature change per fringe shift (Eq. 1.10) shows that the estimated refraction error is

$$\text{refraction error} = -\frac{1}{L\,dn/dT}\frac{1}{6n(y)}\left(\frac{dn}{dT}\right)^2\left(\left(\frac{\partial\overline{T}}{\partial y}\bigg|_2\right)^2 - \left(\frac{\partial\overline{T}}{\partial y}\bigg|_1\right)^2\right)L^3$$

(1.15)

It increases as L^3, namely the cube of the length of the test cell in the viewing direction. It increases further as $(dn/dT)^2$, being considerably higher for liquids in comparison to gases. Refraction error also increases as $(\partial T/\partial y)^2$, the transverse temperature gradient squared and is particularly serious near thermally active surfaces. Thus, interferometry as a quantitative tool loses its effectiveness in problems where heat and mass transfer gradients are quite large. Fringes continue to form and interferometry can still be used for qualitative flow visualization.

Although beam bending is a source of error in interferometry, it forms the basis of measurements in schlieren and shadowgraph—the subject matter of the present monograph [1, 10, 11].

References

1. Goldstein RJ (ed) (1996) Fluid mechanics measurements. Taylor and Francis, New York
2. Hecht J (1986) The laser guidebook. McGraw-Hill, New York
3. Jenkins FA, White HE (2001) Fundamentals of optics. McGraw-Hill, New York
4. Lehner M, Mewes D (1999) Applied optical measurements. Springer, Berlin
5. Lauterborn W, Vogel A (1984) Modern optical techniques in fluid mechanics. Annu Rev Fluid Mech 16:223–244
6. Mayinger F (1993) Image-forming optical techniques in heat transfer: revival by computer-aided data processing. J Heat Transf-Trans ASME 115:824–834
7. Mayinger F (ed) (1994) Optical measurements: techniques and applications. Springer, Berlin
8. Muralidhar K (2001) Temperature field measurement in buoyancy-driven flows using interferometric tomography. Annu Rev Heat Transf 12:265–376
9. Schiebener P, Straub J, Levelt Sengers JMH, Gallagher JS (1990) Refractive index of water and steam as function of wavelength, temperature, and density. J Phys Chem Ref Data 19(3): 677–717
10. Settles GS (2001) Schlieren and shadowgraph techniques. Springer, Berlin, p 376
11. Tropea C, Yarin AL, Foss JF (eds) (2007) Springer handbook of experimental fluid mechanics. Springer, Berlin

Chapter 2
Laser Schlieren and Shadowgraph

Keywords Knife-edge · Gray-scale filter · Cross-correlation ·
Background oriented schlieren.

2.1 Introduction

Schlieren and shadowgraph techniques are introduced in the present chapter. Topics
including optical arrangement, principle of operation, and data analysis are discussed.
Being refractive index-based techniques, schlieren and shadowgraph are to be compared with interferometry, discussed in Chap. 1. Interferometry assumes the passage
of the light beam through the test section to be straight and measurement is based on
phase difference, created by the density field, between the test beam and the reference beam. Beam bending owing to refraction is neglected in interferometry and is a
source of error. Schlieren and shadowgraph dispense with the reference beam, simplifying the measurement process. They exploit refraction effects of the light beam
in the test section. Schlieren image analysis is based on beam deflection (but not
displacement) while shadowgraph accounts for beam deflection as well as displacement [3, 8, 10]. In its original form, shadowgraph traces the path of the light beam
through the test section and can be considered the most general approach among
the three. Quantitative analysis of shadowgraph images can be tedious and, in this
respect, schlieren has emerged as the most popular refractive index-based technique,
combining ease of instrumentation with simplicity of analysis.

P. K. Panigrahi and K. Muralidhar, *Schlieren and Shadowgraph Methods in Heat
and Mass Transfer*, SpringerBriefs in Thermal Engineering and Applied Science,
DOI: 10.1007/978-1-4614-4535-7_2, © The Author(s) 2012

2.2 Laser Schlieren

A basic schlieren setup using concave mirrors that form the letter Z is shown in Fig. 2.1. The Z-type monochrome schlieren system comprises concave mirrors, flat mirrors, a knife-edge, and a laser as a light source. The optical components and the lasers are kept on a common centerline at a certain elevation. Under undisturbed conditions, the original laser beam as well as the center of the collimated beam (after expansion and collimation) fall on the central portion of the optical components. For the images presented in the following chapters, concave mirrors of the schlieren apparatus are of 1.3 m focal length each and 200 mm diameter. Relatively large focal lengths of the concave mirrors make the schlieren technique quite sensitive to the thermal/concentration gradients. Out of the two concave mirrors, the first acts as the collimator while the second concave mirror placed after the test cell decollimates the laser beam with a focus at a knife-edge. Flat mirror M1 directs the diverging laser beam onto the first concave mirror which collimates it into a beam of uniform diameter. The collimated beam passes through the test section and falls on the second concave mirror which focuses it onto the plane of the knife-edge. The test section is placed between the two concave mirrors. Optical elements are supported on adjustable mounts (with permissible movements in x and y directions, z being the direction of propagation of the laser beam). Slight misalignment in the system with respect to the direction of the laser beam can be taken care of by adjusting the mounts. The knife-edge is placed at the focal plane of the second concave mirror. It is positioned to cutoff a part of the light focused on it, so that in the absence of any disturbance in the test section, the illumination on the screen is uniformly reduced. The mount holding the knife-edge ensures that the knife-edge has flexibility in orientation, say, vertical or horizontal as required in the measurement being carried out. The mount permits the movement of the knife-edge parallel to the direction of the laser beam as well as in plane, so as to cutoff the desired extent of light intensity. In practice, the knife-edge is set perpendicular to the direction in which the density gradients are to be observed. In many applications discussed in the present monograph, density gradients are predominantly in the vertical direction and the knife-edge has been kept horizontal. The initial light intensity values (on a gray scale of 0–255) have been chosen to be <20, making the image look uniformly dark.

Schlieren measurements rely on intensity of light measured in turn by the CCD camera. Within limits, the sensor array of the camera is a linear device and converts intensity to voltage in direct proportion. Beyond the range, nonlinearities set in and the camera is said to be saturated. Since the light falling at the knife-edge is a spot, light intensities are large, and a neutral density filter may be used to uniformly reduce illumination reaching the camera. The screen shown in Fig. 2.1 serves this purpose as well. There could, however, be a disadvantage of non-uniformities of the screen creating errors in intensity measurement and should be addressed. Alternatively, an additional lens can collimate the spot formed at the knife-edge and create a light beam of larger diameter and lower intensity for the camera.

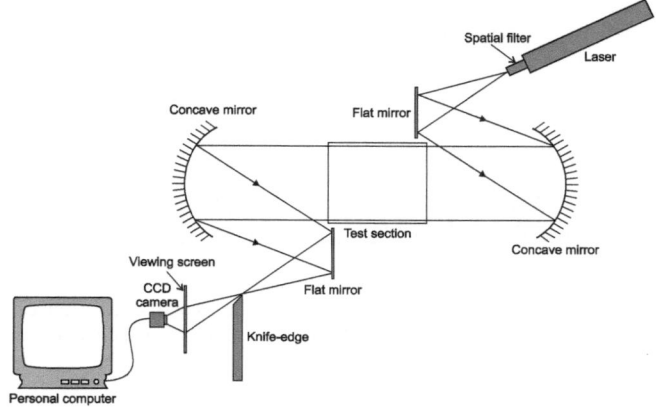

Fig. 2.1 Schematic diagram of a Z-type laser schlieren setup

The issue of camera saturation is serious when lasers are used as a light source. White light sources may also be used in the context of schlieren since coherence is not a matter of concern here. The light source, in the present monograph is, however, a laser, except for discussions related to Chap. 3.

2.2.1 Positioning the Knife-Edge

Before the start of experiments, the schlieren setup (Fig. 2.1) has to be carefully aligned. Apart from issues, such as collimation, focus, and uniform light intensity over the collimated light beam, the adjustment of the knife-edge plays a significant role in the quality of the schlieren image recorded. With best alignment, the undisturbed light beam should form a spot at the knife-edge whose dimensions match the diameter of the pinhole used in the spatial filter (except for scaling introduced by the focal lengths of the optical elements). As the knife-edge is moved to block the spot of light, the intensity of illumination decreases uniformly over the screen. This effect is demonstrated in Fig. 2.2 where the distribution of light intensity over the screen is shown for various positions of the knife-edge. Figure 2.2a and c respectively show non-uniform distribution of light intensity when the knife-edge is either too close or too far away from the second concave mirror. Figure 2.2b shows light intensity distribution for a correctly placed knife-edge partly blocking the light spot. An important step in the alignment procedure is to adjust the percent cutoff by the knife-edge in order to obtain the desired sensitivity. If the cutoff is small, a large amount of light passes over to the screen and results in poor contrast of the schlieren images along with the possibility of camera saturation. If the cutoff is large, high contrast images are possible but the measurement may result in the loss of information in regions of low density gradient.

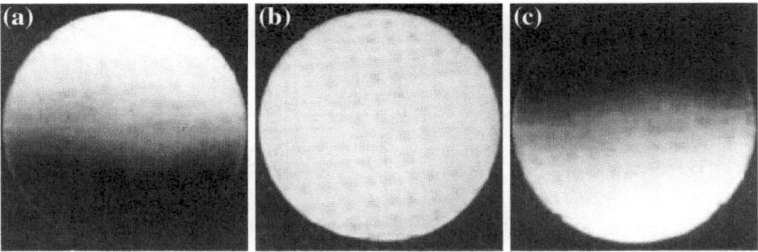

Fig. 2.2 Effect of knife-edge movement on the schlieren image for a horizontal knife-edge cutting off the light beam from below. **a** Knife-edge is very close to the second concave mirror; **b** Knife-edge is correctly placed. **c** Knife-edge is beyond the focal plane of the second concave mirror [8]

Disturbances to a schlieren system include floor vibrations, heavy machinery, and also the movement of laboratory personnel in the vicinity. The schlieren technique, however, is not nearly as vibration-sensitive as interferometry, where motion amplitudes of the order of a light wavelength are visible in the form of fringes. Since schlieren depends primarily on geometric, rather than principles of wave-optics, it is superior to interferometry in its resistance to shock and vibrations. If the sensitivity of the schlieren system is deliberately reduced, either by a lower intensity cutoff by the knife-edge, or by replacing the knife-edge with a graded (gray-scale) filter, vibration errors would be truly minimal.

2.2.2 Analysis of Schlieren Images

The present section analyzes the process of image formation in a schlieren setup. The index of refraction (or its spatial derivatives) determines the resulting light intensity pattern over the screen. An aspect shared by interferometry and schlieren (and indeed, shadowgraph) is that they generate projection data, namely information integrated in the direction of propagation of the light beam. The result is a concentration (or temperature) field that is ray-averaged, specifically, integrated over the length L of the test section [9].

As discussed in Chap. 1, refractive index techniques depend on the unique refractive index—density relationship for transparent media. Called the Lorentz–Lorenz formula, it is expressed as:

$$\frac{n^2 - 1}{\rho(n^2 + 2)} = \text{constant} \tag{2.1}$$

where n is refractive index and ρ, the density. For gases, the refractive index is close to unity and the expression reduces to the Gladstone–Dale equation

$$\frac{n-1}{\rho} = \text{constant} \tag{2.2}$$

For a given wavelength, the constant appearing in Eqs. 2.1 and 2.2 can be evaluated from the knowledge of n and ρ under reference conditions. It depends on the chemical composition of the material and varies slightly with wavelength. In general, density of pure fluids will depend on pressure and temperature. In many applications involving gases, pressure is sensibly constant and density scales entirely with temperature. Liquids such as water are practically incompressible and their density will vary only with temperature. Within limits, this variation of density with temperature can be taken to be linear. Hence, refractive index will itself scale linearly with temperature. For a process involving mass transfer, the Lorentz–Lorenz formula as applied to a solute–solvent system takes the form:

$$\frac{n^2-1}{n^2+2} = \frac{4}{3}\pi(\alpha_A N_A + \alpha_B N_B) \tag{2.3}$$

Here n is the refractive index of the solution, and α and N are respectively the polarizability and mole fraction. This result is often used in crystal growth applications with suffixes A and B specifying water as the solvent and KDP as the solute, respectively [4–6]. The material property that determines the sensitivity of the optical measurement is dn/dT (or dn/dC). Compared to gases, the derivative is about three orders of magnitude larger in liquids. Hence, only a small perturbation is needed to see refraction in liquids compared to air.

Image formation in a schlieren system arises from the deflection of the light beam in a variable refractive index field toward a region of higher refractive index. In order to recover quantitative information from a schlieren image, one has to determine the cumulative angle of refraction of the light beam emerging from the test cell as a function of position in the cross-sectional $x - y$ plane [3]. This plane is normal to the light beam, whose direction of propagation is along the z-direction. The path of the light beam in a medium whose index of refraction varies in the vertical direction (y) can be analyzed using the principles of geometric optics as follows (also see Sect. 1.4.5):

Consider two wavefronts at times τ and $\tau + \Delta\tau$, separated by a small time difference $\Delta\tau$, Fig. 2.3. At time τ the ray is at a position z. After a interval $\Delta\tau$, the light has moved a distance of $\Delta\tau$ times the speed of light c. Since c depends on refractive index, it is a function of y. In addition, the wavefront turns through an angle $\Delta\alpha$. The local speed of light is c_0/n where c_0 is the velocity of light in vacuum and n is the local refractive index of the medium. Hence the distance Δz that the light beam travels during time interval $\Delta\tau$ is

$$\Delta z = \Delta\tau \frac{c_0}{n}$$

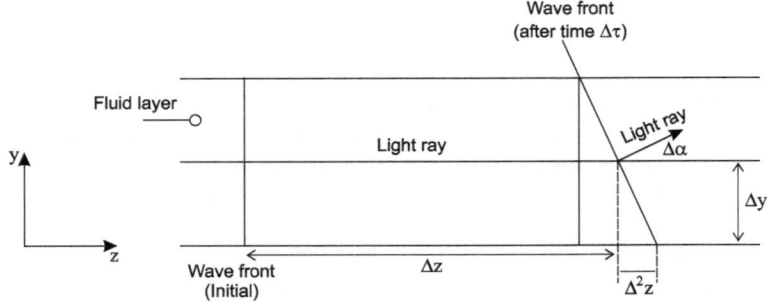

Fig. 2.3 Bending of a light ray in a vertically stratified fluid medium due to refraction

The gradient in refractive index along the y direction results in bending of the wavefront. Let $\Delta\alpha$ represent the bending angle at a location z. A small increment $\Delta\alpha$ in angle α can be expressed as

$$\Delta\alpha \approx \tan(\Delta\alpha) = \frac{\Delta^2 z}{\Delta y}$$

The distance $\Delta^2 z$ is given by

$$\Delta^2 z = \Delta z_y - \Delta z_{y+\Delta y} \approx \Delta z_y - \Delta z_y - \frac{\Delta}{\Delta y}(\Delta z)(\Delta y) = -c_0 \frac{\Delta(1/n)}{\Delta y}\Delta\tau\,\Delta y$$

Hence

$$\Delta\alpha = \frac{\Delta^2 z}{\Delta y} = -c_0 \frac{\Delta(1/n)}{\Delta y}\Delta\tau = -n\Delta z \frac{\Delta(1/n)}{\Delta y}$$

In the limiting case

$$d\alpha = \frac{1}{n}\frac{\partial n}{\partial y}dz = \frac{\partial(\ln n)}{\partial y}dz \tag{2.4}$$

Hence, the cumulative angle through which the light beam has turned over the length L of the test region is

$$\alpha = \int_0^L \frac{\partial(\ln n)}{\partial y}dz \tag{2.5}$$

where the integration is performed over the entire length of the test section. It is to be understood that the angle α is a function of the coordinates x and y on the exit plane of the test cell. If the index of refraction within the test section is different from that of the ambient air (n_a), angle α'' of the light beam emerging from the test cell is given by Snell's law

$$n_a \sin\alpha'' = n \sin\alpha$$

Assuming α and α'' to be small angles, we get

$$\alpha'' = \frac{n}{n_a}\alpha$$

Therefore, from Eq. 2.5, we get

$$\alpha'' = \frac{n}{n_a} \int_0^L \frac{1}{n}\frac{\partial n}{\partial y}dz$$

If the factor $1/n$ within the integrand does not change greatly through the test section, then

$$\alpha'' = \frac{1}{n_a} \int_0^L \frac{\partial n}{\partial y}dz$$

Since $n_a \approx 1.0$ the cumulative angle of refraction of the light beam emerging into the surrounding air is given by

$$\alpha'' = \int_0^L \frac{\partial n}{\partial y}dz \tag{2.6}$$

A schlieren system can be thought of as a device to measure the angle α. In most applications, this angle is quite small, say, of the order of 10^{-6}-10^{-3} radians. It is then reasonable to expect that refractive index gradients lead to the light beam getting displaced in the plane of the knife-e.g., out-of-plane effects being negligible in comparison. The small angle approximation helps in image analysis and is uniformly used through out in the following discussion.

Consider the schlieren measurement system shown in Fig. 2.4 comprising lenses instead of concave mirrors. A light source that generates a light beam of diameter a_s is kept at the focus of lens L_1. Thus, a parallel beam of light is created that probes the density distribution in the test section. The dotted line shows the path of the light beam in the presence of disturbance in the test region. The second lens L_2, whose focus is the knife-edge collects the light beam and passes onto a screen. As discussed by Goldstein [3], the screen is ideally located at the conjugate focus of the test section. This position ensures that intensity changes are related to beam deflection alone, as required in schlieren, and not beam displacement (the *shadowgraph* effect). If no disturbance is present, the passage of the light beam is shown by solid lines reaching the focus of L_2, Fig. 2.5, with diameter a_0. This dimension is related to the initial as

$$\frac{a_0}{a_s} = \frac{f_2}{f_1}$$

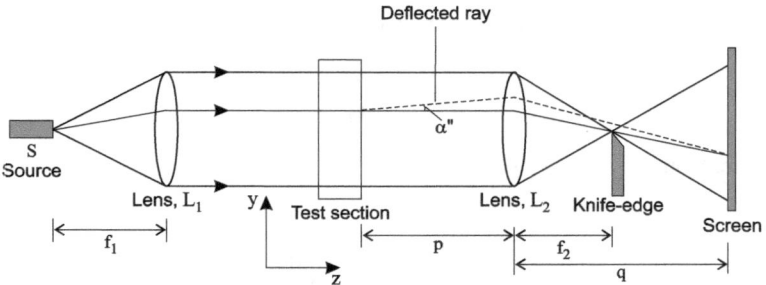

Fig. 2.4 Schematic drawing that shows the path of the light beam in a schlieren system made of lenses. When the screen is at the conjugate focus, the relationship $(1/p) + (1/q) = (1/f_2)$ is followed and the image on the screen is the same size as the cross-section of the test section corresponding to the location p. For the distances shown, the angle detected at the screen is the cumulative turning of the light beam within the test cell. Figure redrawn from [3]

Fig. 2.5 View of undisturbed and deflected light beam cross-sections at the knife-edge of a schlieren system. The horizontal displacement of the light beam does not contribute to intensity contrast

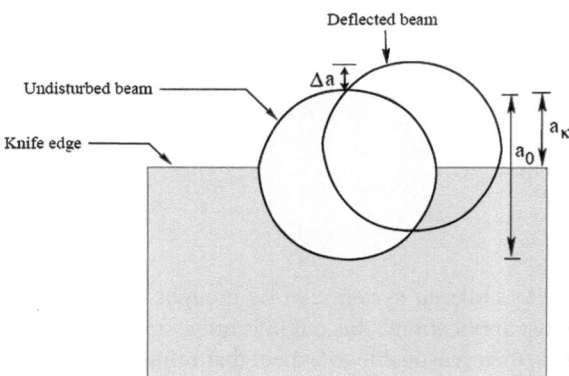

where f_1 and f_2 are the focal lengths of lenses L_1 and L_2, respectively. In a schlieren system, the knife-edge kept at the focal length of the second convex lens is first adjusted, when no disturbance in the test region is present, to cutoff all but an amount corresponding to the dimension a_k of the light beam. Let a_0 be the original size of the laser beam falling on the knife-edge. If the knife-edge is properly positioned, the illumination at the screen changes uniformly, depending upon its direction of the movement. Let I_0 be the illumination at the screen when no knife-edge is present. The illumination I_k with the knife-edge inserted in the focal plane of the second lens but without any disturbance in the test region will be given by

$$I_k = \frac{a_k}{a_0} I_0 \qquad (2.7)$$

Let Δa be the displacement of the light beam in the vertical direction y above the knife-edge corresponding to the angular deflection (α'') of the beam passing through the test region. From Fig. 2.5, Δa can be expressed as

$$\Delta a = \pm f_2 \alpha'' \tag{2.8}$$

The sign in Eq. 2.8 is determined by the direction of the displacement of the laser beam in the vertical direction; it is positive when the shift is in the upward direction and negative if the laser beam gets deflected below the knife-edge. In the following discussion, Eq. 2.8 is considered with a positive sign.

Let I_f be the final illumination on the screen after the light beam has deflected upwards by an amount Δa due to the inhomogeneous distribution of refractive index in the test cell. Hence

$$I_f = I_k \frac{a_k + \Delta a}{a_k} = I_k \left(1 + \frac{\Delta a}{a_k}\right) \tag{2.9}$$

The change in light intensity on the screen ΔI is given by

$$\Delta I = I_f - I_k$$

The contrast thus generated by the schlieren measurement is expressed as

$$\text{contrast} = \frac{\Delta I}{I_k} = \frac{I_f - I_k}{I_k} = \frac{\Delta a}{a_k} \tag{2.10}$$

Using Eq. 2.8

$$\text{contrast} = \frac{\Delta I}{I_k} = \frac{\alpha'' f_2}{a_k} \tag{2.11}$$

Equation 2.11 shows that the contrast in a schlieren system is directly proportional to the focal length of the second concave mirror i.e. f_2. Larger the focal length, greater will be the sensitivity of the system.

Combining Eqs. 2.6 and 2.11, the governing equation in a schlieren system is obtained as

$$\frac{\Delta I}{I_k} = \frac{f_2}{a_k} \int_0^L \frac{\partial n}{\partial y} dz \tag{2.12}$$

Equation 2.12 shows that the schlieren technique records the *path integrated* gradient of refractive index over the length of the test section. If the field is 2D (in the $x - y$) plane, the quantity $\partial n / \partial y$ is independent of the z coordinate and

$$\frac{\Delta I}{I_k} = \frac{f_2}{a_k} \frac{\partial n}{\partial y} L \tag{2.13}$$

The quantity on the left-hand side can be obtained by using the initial and final intensity distributions on the screen. In the experiments discussed in the present monograph, the knife-edge is adjusted in such a way that it cuts off approximately 50 % of the original light intensity, i.e. $a_k = a_0/2$ where a_0 is the original dimension of the laser beam at the knife-edge. The exact value of a_0 cannot be measured. Its value is of the order of microns and can be confirmed only by validation against benchmark experiments. With $a_k = a_0/2$, we get

$$\frac{\Delta I}{I_k} = \frac{2 f_2}{a_0} \frac{\partial n}{\partial y} L \tag{2.14}$$

Equation 2.14 represents the governing equation for the schlieren process in terms of the ray-averaged refractive index field. Since ΔI is calculated purely in terms of the angle α, the model presented above requires that changes in light intensity occur due to beam deflection alone, rather than its physical displacement.

The above derivation can be repeated for a knife-edge held vertical so that x-derivatives of refractive index can be imaged on the screen. This approach makes the *paraxial* approximation in that the derivatives in x and y directions are taken to have independent influences on beam deflection. It is expected to hold under the small angle approximation adopted in this chapter.

If the working fluid is a gas, the first derivative of the refractive index field with respect to y can be expressed in terms of density using Eq. 2.2 as

$$\frac{\partial \rho}{\partial y} = \frac{\rho_0}{n_0 - 1} \frac{\partial n}{\partial y} \tag{2.15}$$

Equation (2.15) relates the gradient in the refractive index field with the gradients of the density field in the fluid medium inside the test cell. The governing equation for schlieren measurement in gas can be rewritten as

$$\frac{\Delta I}{I_k} = \frac{f_2}{a_k} \frac{n_0 - 1}{\rho_0} \frac{\partial \rho}{\partial y} L \tag{2.16}$$

Assuming that the pressure inside the test cell is practically constant, we get

$$\frac{\Delta I}{I_k} = -\frac{f_2}{a_k} \frac{n_0 - 1}{\rho_0} \frac{p}{RT^2} \frac{\partial T}{\partial y} L \tag{2.17}$$

Equations 2.16 and 2.17 respectively relate the contrast measured using a laser schlieren technique with the density and temperature gradients in the test section. With the dependent variables such as T defined away from a solid surface or with proper boundary conditions, these equations can be integrated to determine the quantity of interest. For a KDP solution arising in crystal growth applications [4–6], Eq. 2.3 governs the relationship between species concentration (expressed as a mole fraction N) and refractive index. The concentration gradient is now obtained as

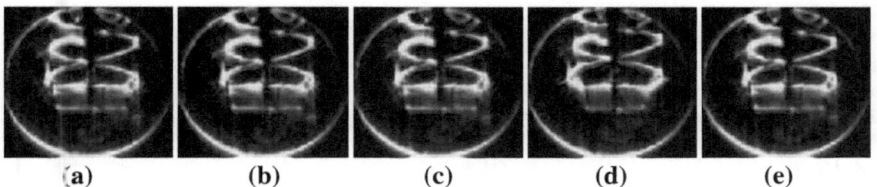

(a)   (b)   (c)   (d)   (e)

Fig. 2.6 Original schlieren images **a–d** of convective field as recorded by the CCD camera and the corresponding time-averaged image **e**, from [9]

$$\frac{\partial N}{\partial y} = \frac{9n}{2\alpha_{KDP}(n^2 + 2)^2} \frac{\partial n}{\partial y} \qquad (2.18)$$

where α_{KDP} is the polarizability of the KDP material in water ($=4.0\,cm^3$/mole) and N is the molar concentration of the solution. Combining Eqs. 2.12 and 2.18 and integrating from a location in the bulk of the solution (where the gradients are negligible), the concentration distribution around the growing crystal can be uniquely determined.

Equations 2.16 and 2.17 show that the schlieren measurements are primarily based on the intensity distribution as recorded by the CCD camera. Since calculations are based on an intensity ratio, it is not necessary to record absolute intensity values. This step requires the camera to be a linear device that converts intensity to voltages stored in the computer. The linearity requirement can be fulfilled by ensuring that the camera does not get saturated with light intensity. A second implication of Eq. 2.14 is that original light intensities are required and schlieren images should not be subject to image processing operations. In practice, the camera sensors may show pixel-level scatter and it is common to average intensities over a 3×3 pixel template. In fluids, a perfectly steady convective field may not be attained and temporal fluctuations are possible. In such instances, a certain amount of time-averaging is performed before starting data analysis. Figure 2.6 shows one such set of four schlieren images recorded as a time sequence and their averaged image. The images show the convective plume in the form of high intensity regions above a crystal growing from its aqueous solution and are discussed in detail in [9].

2.2.2.1 Window Correction

For imaging the temperature or concentration field by laser schlieren, optical windows are often used to contain the fluid region within. Such windows must be used while working with liquids, but may be required in gases as well if the influence of the environment is to be minimized. The optical windows employed are of finite thickness (say, 5 mm) and the index of refraction of its material (for example, BK-7) is considerably different from that of the liquid within and the ambient air. The light beam emerging out of the test section with an angular deflection undergoes refraction

Fig. 2.7 Schematic drawing
of the path of the light beam
and the corresponding angles
of deflection as it passes
through the growth chamber.
Angles shown are exaggerated
for clarity

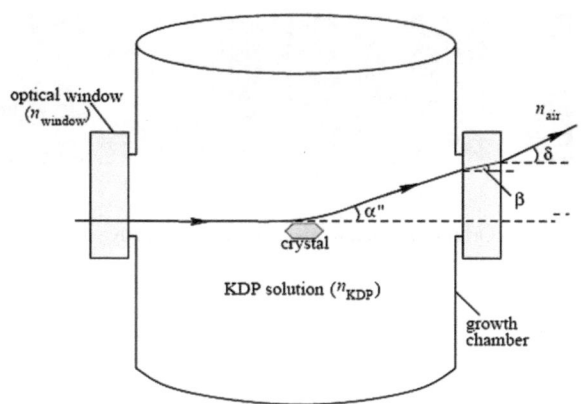

again before falling on the second concave mirror. The contribution of refraction at
the confining optical windows can be accounted for by applying a correction factor
to Eq. 2.14 as discussed below.

Consider the growth of a KDP crystal from its supersaturated solution in water
as shown in Fig. 2.7. The laser beam strikes the second optical window fixed on the
growth chamber at an angle after undergoing refraction due to variable concentration
gradients in the vicinity of the growing KDP crystal. The optical windows have an
index of refraction equal to n_{window} (around 1.509). The refractive index of the KDP
solution (n_{KDP}) at an average temperature of 30 °C is 1.355, and for air n_{air} is close
to unity. Let α'' be the angular deflection of the beam purely due to the presence of
concentration gradients in the vicinity of the growing crystal, Fig. 2.7. The beam
strikes the second optical window at this angle. Let β be the angle at which light
leaves the inner surface of the second optical window. Using Snells law, we get

$$\frac{n_{\text{KDP}}}{n_{\text{window}}} = \frac{\sin \beta}{\sin \alpha''} \tag{2.19}$$

Since α'' is quite small in most applications, $\sin \alpha'' \approx \alpha''$ and

$$\sin \beta \approx \beta = \left(\frac{n_{\text{KDP}}}{n_{\text{window}}} \right) \alpha'' \tag{2.20}$$

Let δ be the final angle of refraction with which the laser beam emerges into the
surrounding air. For the optical window-air combination

$$\left(\frac{n_{\text{window}}}{n_{\text{air}}} \right) = \frac{\sin \delta}{\sin \beta} \tag{2.21}$$

Hence

$$\sin \delta = \left(\frac{n_{\text{window}}}{n_{\text{air}}} \right) = \left(\frac{n_{\text{KDP}}}{n_{\text{window}}} \right) \alpha'' \tag{2.22}$$

or

$$\sin \delta \approx \delta = \left(\frac{n_{\text{KDP}}}{n_{\text{air}}} \right) \alpha'' \tag{2.23}$$

In experiments with optical windows, schlieren image analysis discussed in the previous sections would have to be carried out by first computing α'' from the recorded angle δ.

2.2.3 Gray-Scale Filter

The knife-edge of a conventional schlieren system is an excellent device for improving light intensity contrast in the optical image, but suffers from certain drawbacks. For example, gradients parallel to the knife-edge do not contribute to image formation. When the light beam is deflected below the knife-e.g., the gradient information is lost. Higher order effects such as a focus formed beyond the knife-edge may arise in measurements. Intensity modulation can also occur from diffraction of light at the sharp e.g., resulting in stray interference patterns superimposed on the schlieren image. Many of these drawbacks can be addressed by using a gray-scale (*graded*) filter. The filter is a photographic film on which a computer generated gray-scale is printed. The filter width may match that of the knife-edge. The vertical extent of the filter can be tuned to the deflections of light anticipated on the filter plane. The grayscale values of light intensity may vary from 0 to 255 (for a camera with 8-bit resolution) or a part of the range. The initial setting of the filter with respect to the undisturbed light spot is also an adjustable quantity. If the spot falls at the center of the filter, positive as well as negative beam deflections can be determined. The knife-edge can be thought of as a special construction of a gray-scale filter with two shades of 0 and 255 with a sharp cutoff.

In schlieren measurements with lenses, the diverging light beam from the spot formed over the knife-edge falls on a screen and the image recorded by the camera. In a Z-type configuration, the camera may be focussed on the light spot falling on the knife-edge. However, such an arrangement can lead to camera saturation. It is preferable to allow image formation on a screen and record the image at parallel incidence. In a graded filter arrangement, the filter itself acts as the screen and the camera records the image directly from it. In this approach, the filter needs to be calibrated for light intensity as a function of beam displacement at the filter location. This step is conveniently carried out when the test cell is undisturbed and the filter, mounted on a vertical traverse, moves relative to the spot of light. Under test conditions, the change in intensity at a point is mapped to beam displacement via the calibration curve. When a white light source such as a xenon lamp is used, an additional lens may be used to collimate light from the spot formed at the filter plane and deliver it to the CCD camera.

Variations in the absorptivity of the photographic film (or the material used as a graded filter) can influence intensity measurement. An alternative is to use a color filter along with a color CCD camera. Here, color measured in terms of hue, dispenses with intensity, and material imperfections do not give rise to additional errors in measurements. A color filter used instead of the gray-scale, generates color images of the convective field. This approach, called *rainbow schlieren* is discussed in Chap. 3.

2.3 Background Oriented Schlieren

Background oriented schlieren (BOS) is a technique in which image variations of a distinct background are analyzed to determine density variations in a flow field. The index of refraction of a transparent medium has a direct correspondence to the density of the fluid. Therefore, density gradients cause index of refraction gradients. Rays of light passing through a test section are bent, to an extent that depends on the density gradient in the experimental test cell. This process alters the perception of the background image. The dependence of image formation on the refractive index field in BOS is similar to the basic schlieren setup, but BOS can be implemented in a much simpler apparatus as discussed below.

2.3.1 Experimental Details

The schematic drawing of a BOS setup is shown in Fig. 2.8. While the basic schlieren configuration often needs several high-quality lenses and mirrors guide the light beam, BOS needs only an illuminated background image, a CCD camera, and a computer with image acquisition software. The absence of precision optical components makes BOS a cheaper alternative. It also allows BOS to be more easily scalable to whatever size and precision is needed to accurately capture the density field in and around a given test model. A classical schlieren is preferably operated in a darkroom environment, since any ambient light can contaminate the image. In contrast, a BOS can operate with additional light sources, as the BOS technique is based on the virtual displacement of the background image, not just the intensity of light reaching the camera. In order for BOS to generate meaningful images, the background image must have high contrast and must be sensitive to small displacements. A randomized grid of small black dots on a white background serves this purpose well.

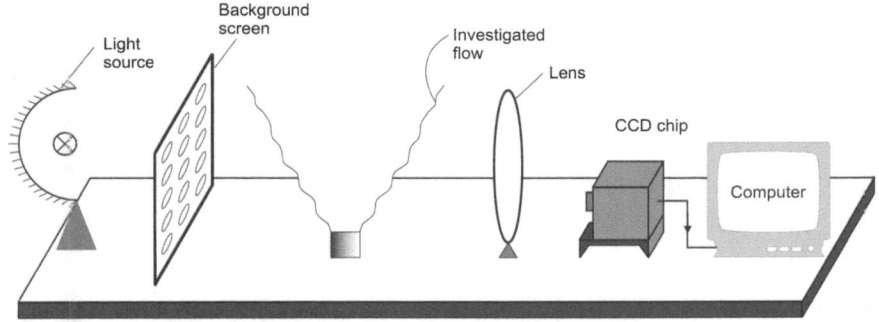

Fig. 2.8 Schematic diagram of the background oriented schlieren setup

2.3.2 Data Analysis

Figure 2.9 is a schematic drawing that explains the principle behind imaging by a BOS system. Here, z represents the coordinate along the light path, f is the focal length of the camera lens, Z_C is the distance from the camera to the phase object field, and Z_B is the distance from the phase object to the background image. The local image displacement χ can be expressed by integrating the local refractive index gradients along the light path as:

$$\chi = \frac{f Z_B}{Z_C + Z_B - f} \int_{\Delta_z} \frac{1}{n_0} \frac{\partial n}{\partial r} dz, \tag{2.24}$$

Here, n, refractive index field is a function of the cross-sectional plane coordinates (x, y). The 2D image displacement value $\chi(x, y)$ can be used to determine the partial derivatives $\partial \rho / \partial x$ and $\partial \rho / \partial y$ by using Lorentz–Lorenz relationship, as in image analysis used for the basic schlieren arrangement. BOS employs a computer generated dot pattern screen, placed behind the test cell. The object field (namely, the test cell) defined by its density variation is placed between the camera and the dot pattern. For BOS, two image sets are recorded. One dot pattern image is acquired without density effects. The second dot pattern image is acquired with the density gradient prevailing in the test cell. The displacements of the dots are calculated using an image displacement correlation algorithm. The software often used for this type of processing is readily adapted from particle image velocimetry (PIV), which is commonly used experimental fluid mechanics laboratories. The initial image may carry uniformly distributed dots or otherwise, and may be tailored to the application being studied. The second image can be a time sequence of images, if the phenomenon of interest is unsteady. While the displacement of particles in a series of PIV images corresponds to velocities, the displacement of background dots in BOS images correspond to the density variation. The spatial resolution of measurement is determined by the dot size. As opposed to basic schlieren, BOS measures, not small

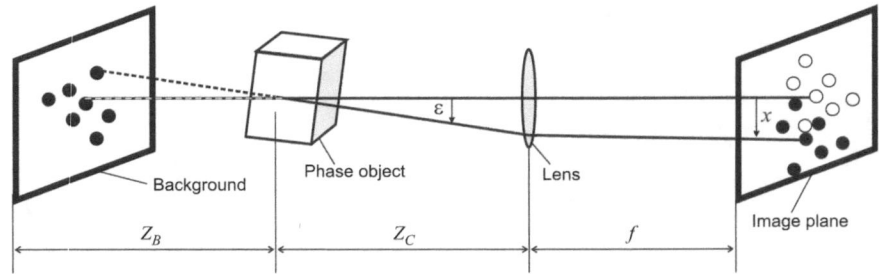

Fig. 2.9 Schematic drawing that explains the image formation in background oriented schlieren

angles but small displacements, and can be advantageous in certain contexts. Beam displacement errors (related to the shadowgraph effect) need to be accounted for during image analysis. An additional set of references on BOS is provided at the end of the chapter.

The cross-correlation algorithm used for determining displacement is shown schematically in Fig. 2.10. Let I_1 and I_2 be the interrogation regions of the initial and the final images being cross-correlated. The images are defined in terms of intensities at pixel indices (i, j) with the pixel sizes being Δx and Δy in the two directions. These indices run over $i = 1 \ldots M$ and $j = 1 \ldots M$. The cross-correlation function R_{12} between this pair of images is numerically evaluated as

$$R_{1,2}(i, j) = \sum_{l=1}^{M} \sum_{m=1}^{N} I_1(l, m) I_2(l + i - 1, m + j - 1), \quad i = 1 \ldots M; \quad j = 1 \ldots N$$

In practice, the cross-correlation function is evaluated using Fourier transforms to exploit the efficiency of the fast Fourier transform (FFT) algorithm. Let the 2D Fourier transforms of these images be respectively given as $\hat{I}_1$ and $\hat{I}_2$, while $*$ indicates complex conjugate. The symbol IFT is used for the inverse Fourier transform of its argument. In terms of Fourier transforms, the cross-correlation function is written as

$$R_{12} = \mathrm{IFT}\{\hat{I}_1 \times \hat{I}_2^{*}\}$$

Here, IFT is inverse Fourier transform and can also be evaluated using the FFT algorithm. Such calculations can be carried out using commercially available image analysis software. Displacement information associated with the interrogation spot is contained in the spatial coordinates (in integer multiples of the pixel size Δx and Δy) where the cross-correlation function attains its maximum. Displacement can now be related to beam deflection and hence, refractive index gradients prevalent in the physical domain.

Fig. 2.10 Pictorial representation of the cross-correlation algorithm for displacement calculation

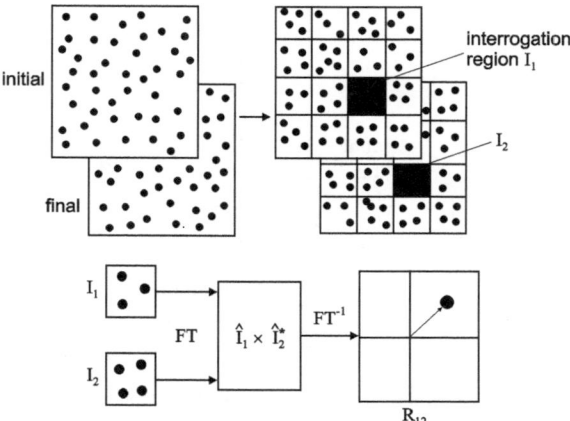

2.4 Shadowgraph Technique

Shadowgraph has been extensively used in experimental fluid mechanics and heat transfer but as a tool for flow visualization. Shadowgraph employs an expanded collimated beam of light from a laser that traverses the field of disturbance. If the disturbance is one of varying refractive index, the individual light rays passing through the test section are refracted and bent out of their original path. This causes a spatial modulation of the light-intensity distribution with respect to the original intensity on the screen. The resulting pattern is a shadow of the refractive index field prevailing in the region of the disturbance. Figure 2.11 shows the schematic drawing of the shadowgraph arrangement [11]. A He–Ne laser (15–35 mW power, continuous wave) is expanded and collimated to a suitable diameter by a beam expander. The collimated beam passes through the test section being investigated. The beam emerging from the exit window falls on a screen resulting in the shadowgraph image. The images may be recorded as individual frames or a video sequence by a suitable camera. A sample shadowgraph of a slightly heated water jet is shown in Fig. 2.12. Here, the initial instability as well as the breakdown of ring vortices to small-scale turbulence are visible.

The discussion related to intensity distortions by a screen in the context of schlieren carries over to the shadowgraph. With a laser as a light source, the screen can diminish light intensity and prevent camera saturation. When a distributed light source is used, the screen can be replaced by a lens arrangement that conveys all the available light to the CCD array. The advantage here is that the camera can focus on any plane beyond the test cell and the sensitivity of measurement suitable altered.

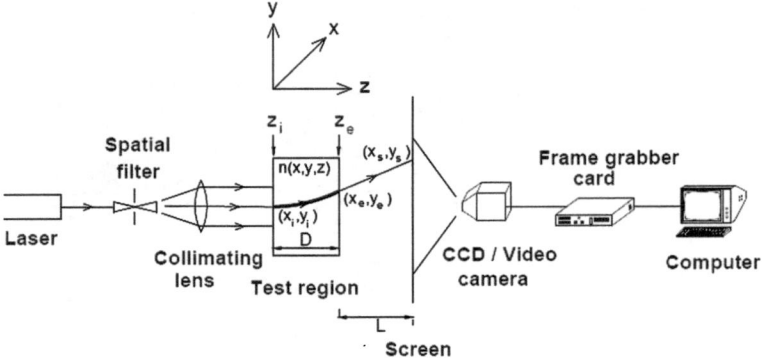

Fig. 2.11 Schematic drawing of the shadowgraph technique

Fig. 2.12 a Initial image of the laser beam before flow is turned on. **b** Shadowgraph of a slightly heated water jet at a Reynolds number of 693 based on nozzle diameter and averaged fluid speed. Ring vortices initially formed and their progressive breakdown to turbulent structures are visible

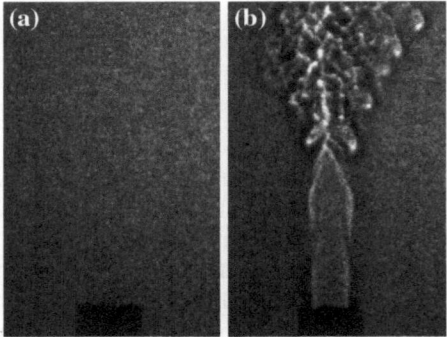

2.4.1 Governing Equation and Approximations

Quantitative data can be extracted from shadowgraph images using the formulation discussed below [7, 11].

Consider a medium with refractive index n that depends on all the three space coordinates, namely $n = n(x, y, z)$. We are interested in tracing the path of light rays as they pass through this region. Starting with the knowledge of the angle and the point of incidence of the ray at the entrance plane, we would like to know the location of the point on the exit window, and the slope of the emergent ray [1, 7]. Let the ray be incident at point $p_i(x_i, y_i, z_i)$ while the exit point is $p_e(x_e, y_e, z_e)$. According to Fermat's principle, the optical path length (OPL) traversed by the light beam between these two points has to be an extremum. For a geometric segment ds along the path, the corresponding OPL is $n(x, y, z)ds$. The total OPL between the two points p_i and p_e is

$$\text{OPL} = \int_{p_i}^{p_e} n(x, y, z) ds$$

As per Fermat's principle

$$\delta \left(\int_{p_i}^{p_e} n(x, y, z) ds \right) = 0 \qquad (2.25)$$

Parameterizing the light path by z, the ray is fully defined by functions $x(z)$ and $y(z)$. Equation 2.25 can be rewritten as

$$\delta \left(\int_{z_i}^{z_e} n(x, y, z) \sqrt{x'^2 + y'^2 + 1} \, dz \right) = 0 \qquad (2.26)$$

where the primes denote differentiation with respect to z. Application of the variational principle to Eq. 2.26 generates the following two, coupled, Euler–Lagrange equations:

$$x''(z) = \frac{1}{n}(1 + x'^2 + y'^2) \left(\frac{\partial n}{\partial x} - x' \frac{\partial n}{\partial z} \right) \qquad (2.27)$$

$$y''(z) = \frac{1}{n}(1 + x'^2 + y'^2) \left(\frac{\partial n}{\partial y} - y' \frac{\partial n}{\partial z} \right) \qquad (2.28)$$

Here (for x)

$$x'(z) = \frac{dx}{dz} \text{ and } x''(z) = \frac{d^2 x}{dz^2}$$

Equations 2.27 and 2.28 are second-order nonlinear ordinary differential equations for $x(z)$ and $y(z)$. The four constants of integration required to solve these equations come from the boundary conditions at the entry plane of the chamber. These are the co-ordinates $x_i = x(z_i)$ and $y_i = y(z_i)$ of the entry point as well as the local derivatives $x_i' = x'(z_i)$ and $y_i' = y'(z_i)$. The solution of the two differential equations yields two orthogonal components of the deflection of the light ray at the exit plane along with its slope and curvature. In most experiments, the light beam entering the apparatus is parallel (at normal incidence) and the respective slopes are zero, namely $x_i' = 0$ and $y_i' = 0$. Let the length of the experimental chamber containing the fluid be D and the screen located at a distance L beyond the exit plane. The z-coordinates of individual light rays at entry, exit, and on the screen are given by z_i, z_e and z_s respectively. Since the incident beam is normal to the entrance plane, there is no refraction at the first optical window. While the derivatives of all the incoming light rays at the entry plane are zero, the displacements $x_s - x_i$ and $y_s - y_i$ of a light ray on the screen (x_s, y_s) with respect to its entry position (x_i, y_i) are

$$x_s - x_i = (x_e - x_i) + L \times x'(z_e) \qquad (2.29)$$

$$y_s - y_i = (y_e - y_i) + L \times y'(z_e) \qquad (2.30)$$

Here, the coordinates x_e, y_e, and $x'(z_e)$, $y'(z_e)$ are given by the solutions of Eqs. 2.27 and 2.28. The first term accounts for refraction within the physical domain, while the second term is the passage of light in normal ambient, taken to be undisturbed, along a straight line. These equations define image formation on the screen in a shadowgraph process and provide a route toward indirect determination of refractive index distribution from the Euler–Lagrange equations.

The formulation given above can be simplified under the following assumptions:

Assumption 2.1 Assume that the light rays at normal incidence on the entrance plane undergo only infinitesimal deviations inside the inhomogeneous field, but have a finite curvature on leaving the experimental apparatus. The derivatives $x'(z_i)$ and $y'(z_i)$ are zero, whereas $x'(z_e)$ and $y'(z_e)$ at the exit plane are finite values. The assumption is justified in contexts where the medium is weakly refracting. Equations 2.27– 2.30 can now be simplified as

$$x''(z) = \frac{1}{n}\left(\frac{\partial n}{\partial x}\right) \tag{2.31}$$

$$y''(z) = \frac{1}{n}\left(\frac{\partial n}{\partial y}\right) \tag{2.32}$$

$$x_s - x_i = Lx'(z_e) \tag{2.33}$$

$$y_s - y_i = Ly'(z_e) \tag{2.34}$$

Rewriting Eqs. 2.33 and 2.34 as

$$x_s - x_i = L\int_{z_i}^{z_e} x''(z)\mathrm{d}z \tag{2.35}$$

$$y_s - y_i = L\int_{z_i}^{z_e} y''(z)\mathrm{d}z \tag{2.36}$$

and using Eqs. 2.31 and 2.32, Eqs. 2.35 and 2.36 become

$$x_s - x_i = L\int_{z_i}^{z_e} \frac{\partial(\log n)}{\partial x}\mathrm{d}z \tag{2.37}$$

$$y_s - y_i = L\int_{z_i}^{z_e} \frac{\partial(\log n)}{\partial y}\mathrm{d}z \tag{2.38}$$

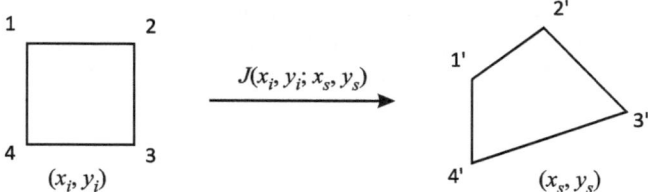

Fig. 2.13 Jacobian $J(x_i, y_i; x_s, y_s)$ of the mapping function connecting the original area (x_i, y_i) to the deformed area (x_s, y_s)

Note that the ray displacements are obtained as an *integration* over the length of the apparatus, and are *path integrals* in this respect.

Assumption 2.2 The assumption of infinitesimal displacement inside the growth chamber can be extended and taken to be valid even for the region falling between the screen and the exit plane of the chamber. As a result, the coordinates of the ray on the screen can be written as

$$x_s = x_i + \delta_x(x_i, y_i) \tag{2.39}$$
$$y_s = y_i + \delta_y(x_i, y_i) \tag{2.40}$$

The deviation of light rays from their original path in the physical medium results in a change of intensity distribution on the screen as compared to the original, when the physical region is undisturbed. Shadowgraph measures this change in the intensity distribution and relates it to the refractive index distribution. The intensity at point (x_s, y_s) on the screen is a result of several beams moving from locations (x_i, y_i) and getting mapped onto the point x_x, y_s on the screen. Since the initial spread of the light beam is deformed during its passage through the apparatus, the light intensity at point (x_s, y_s) is

$$I_s(x_s, y_s) = \sum_{(x_i, y_i)} \frac{I_0(x_i, y_i)}{\left| \frac{\partial(x_s, y_s)}{\partial(x_i, y_i)} \right|} \tag{2.41}$$

Here I_s is the intensity on the screen in the presence of an inhomogeneous refractive index field, and I_o is the original undisturbed intensity distribution. The denominator in the above equation is the Jacobian $J(x_i, y_i; x_s, y_s)$ of the mapping function that connects (x_i, y_i) with (x_s, y_s), as shown in the Fig. 2.13. Geometrically it represents the ratio of the area enclosed by four adjacent rays after and before passing through the test section. In the absence of any disturbance, a small rectangle maps onto a an identical rectangle of equal area and the Jacobian is unity. The summation in the above equation extends over all the rays passing through points (x_i, y_i) at the entry of the test section that are mapped onto the small quadrilateral (x_s, y_s) on the screen and contribute to the light intensity within.

Assumption 2.3 Under the assumption of infinitesimal displacements, the deflections δ_x and δ_y are small. Therefore, the Jacobian can be assumed to be linearly dependent on them. Neglecting higher powers of δ_x and δ_y and also their product, Jacobian can be expressed as

$$\left| \frac{\partial(x_s, y_s)}{\partial(x_i, y_i)} \right| \approx 1 + \frac{\partial(x_s - x_i)}{\partial x} + \frac{\partial(y_s - y_i)}{\partial y} \tag{2.42}$$

Substituting in Eq. 2.41, we get

$$I, (x_s, y_s) \left[1 + \frac{\partial(x_s - x_i)}{\partial x} + \frac{\partial(y_s - y_i)}{\partial y} \right] = \sum_{(x_i, y_i)} I_0(x_i, y_i) \tag{2.43}$$

Simplifying further

$$\frac{I_0(x_i, y_i) - I_s(x_s, y_s)}{I_s(x_s, y_s)} = \frac{\partial(x_s - x_i)}{\partial x} + \frac{\partial(y_s - y_i)}{\partial y} \tag{2.44}$$

Using Eqs. 2.37 and 2.38 for $(x_s - s_i)$ and $(y_s - y_i)$ and integrating over the dimensions of the experimental apparatus, we get

$$\frac{I_0(x_i, y_i) - I_s(x_s, y_s)}{I_s(x_s, y_s)} = (L \times D) \left(\frac{\partial^2}{\partial x^2} + \frac{\partial^2}{\partial y^2} \right) \log n(x, y) \tag{2.45}$$

Here, $n(x, y)$ is to be interpreted as an average value of refractive index over a length L in the z-direction. Equation 2.45 is the governing equation of the shadowgraph process under the set of linearizing approximations 1–3. In concise form the above equation can be rewritten as

$$\frac{I_0 - I_s}{I_s} = (L \times D) \nabla^2 \log n(x, y) \tag{2.46}$$

2.4.2 Numerical Solution of the Poisson Equation

The linearized governing differential equation of the shadowgraph process (Eq. 2.46) is a *Poisson* equation. In measurements, the left side of this equation is recorded as the shadowgraph image. The Poisson equation relates light intensity variation in the shadowgraph image to the refractive index field of the physical medium. In order to solve for refractive index, the following numerical procedure can be adopted. First, the Poisson equation is discretized over the physical domain of interest by a finite-difference method. The resulting system of algebraic equations is solved for the shadowgraph image under consideration to yield a depth-averaged refractive index value for each node of the grid. A mix of Dirichlet and Neumann conditions are usually available to serve as boundary conditions. This approach is considerably simpler than solving the inverse problem indicated in Eqs. 2.27 and 2.28.

In order to assess the validity of assumptions 1–3, the importance of higher order optical effects in shadowgraph imaging need to be examined. This step is accomplished by determining the extent of bending of light rays in a field of known refractive index. A possible approach is to solve the Poisson equation for the refractive index field and then evaluate ray displacements from Eqs. 2.27 and 2.28 where refractive index appears as a parameter. A useful guideline for linearity to hold is that the Jacobian, constructed using four adjacent points and interpreted as per Fig. 2.13 remains within ±5 % of unity.

2.5 Closure

Image formation in interferometry, schlieren, and shadowgraph relies on refractive index changes in the physical domain. Interferometry has a larger number of optical elements. Since it is based on differential measurement of phase, it is sensitive to alignment. Schlieren has fewer optical components and is less sensitive and shadowgraph, being the simplest configuration, is the least sensitive to factors, such as alignment, vibrations, and other extraneous factors. Interferograms are quite vivid, since fringes are isotherms (iso-concentration lines) and unambiguously represent the temperature (concentration) field. The discusson in Chap. 1 shows that the analysis of interferograms is quite straightforward. Schlieren and shadowgraph images reveal regions of high concentration gradients in the form of heightened (or diminished) brightness. Temperature and concentration can be recovered in schlieren by integrating the intensity distribution. In shadowgraph, a Poisson equation needs to be solved, subject to suitable boundary conditions. Thus, a shadowgraph experiment is the easiest to perform while the analysis of shadowgraph data is the most complicated. Schlieren, in this respect, falls between interferometry and shadowgraph with modest demands on experimental complexity and data reduction.

The three refractive index-based techniques yield images that are integrated values of temperature/concentration (or their derivative in a cross-sectional plane) in the direction of propagation of the light beam. If the spatial extent of the disturbed zone in the domain is small, the information contained in the image is small. In the context of interferometry, the consequence could be the appearance of too few fringes in the infinite fringe setting and small fringe deformation in the wedge fringe setting. In schlieren and shadowgraph, weak disturbances show up as small changes in intensity and hence, contrast. The difficulty can be alleviated in schlieren by using large focal length optics so that small deflections are amplified. In shadowgraph, image quality can be improved by moving the screen away from the test cell. The sensitivity of interferometric measurements can be improved by using techniques such as phase shifting but they require additional optical components and revised analysis tools. Additional difficulties with interferometry are the need for maintaining identical experimental conditions in the test section and the compensation chamber, careful balancing of the test and the reference beams, and limitations arising from the fact that quantitative information is localized at the fringes.

This discussion shows that configuring the interferometer as an instrument for process control poses the greatest challenge, schlieren, and shadowgraph being relatively simpler. Schlieren may be considered as an optimum while comparing the ease of analysis with the difficulty of instrumentation.

References

1. Born M, Wolf E (1980) Principles of optics. Pergamon Press, Oxford
2. Gebhart B, Jaluria Y, Mahajan RL, Sammakia B (1988) Buoyancy-induced flows and transport. Hemisphere Publishing Corporation, New York
3. Goldstein RJ (ed) (1996) Fluid mechanics measurements. Taylor and Francis, New York
4. Mantani M, Sugiyama M, Ogawa T (1991) Electronic measurement of concentration gradient around a crystal growing from a solution by using Mach-Zehnder interferometer. J Cryst Growth 114:71–76
5. Onuma K, Tsukamoto K, Nakadate S (1993) Application of real time phase shift interferometer to the measurement of concentration field. J Cryst Growth 129:706–718
6. Rashkovich LN (1991) KDP family of crystals. Adam Hilger, New York
7. Schopf W, Patterson JC, Brooker AMH (1996) Evaluation of the shadowgraph method for the convective flow in a side-heated cavity. Exp Fluids 21:331–340
8. Settles GS (2001) Schlieren and shadowgraph techniques. Springer, Berlin, p 376
9. Srivastava A (2005) Optical imaging and control of convection around a KDP crystal growing from its aqueous solution, Ph.D. thesis, IIT Kanpur (India)
10. Tropea C, Yarin AL, Foss JF (eds) (2007) Springer handbook of experimental fluid mechanics. Springer, Berlin
11. Verma S (2007) Convection, concentration and surface feature analysis during crystal growth from solution using shadowgraphy, interferometry and tomography, Ph.D. thesis, IIT Kanpur (India)
12. Atcheson B, Heidrich W, Ihrke I (2009) An evaluation of optical flow algorithms for background oriented schlieren imaging. Exp. in Fluids 46:467–476
13. Goldhahn E, Seume J (2007) The background oriented schlieren technique: sensitivity, accuracy, resolution and application to a three-dimensional density field. Exp. in Fluids 43:241–249
14. Kindler K, Goldhahn E, Leopold F, Raffel M (2007) Recent developments in background oriented Schlieren methods for rotor blade tip vortex measurements. Exp. Fluids 43:233–240
15. Ramanah D, Raghunath S, Mee DJ, Rsgen T, Jacobs PA (2007) Background oriented schlieren for flow visualisation in hypersonic impulse facilities. Shock Waves 17:65–70
16. Roosenboom EWM, Schroder A (2009) Qualitative Investigation of a Propeller Slipstream with Background Oriented Schlieren. Journal of Visualization 12(2):165–172
17. Sommersel OK, Bjerketvedt D, Christensen SO, Krest O, Vaagsaether K (2008) Application of background oriented schlieren for quantitative measurements of shock waves from explosions. Shock Waves 18:291–297
18. Sourgen F, Leopold F, Klatt D (2012) Reconstruction of the density field using the Colored Background Oriented Schlieren Technique(CBOS). Optics and Lasers in Engg 50:29–38
19. Venkatakrishnan L, Meier GEA (2004) Density measurements using the Background Oriented Schlieren technique. Exp. in Fluids 37:237–247

Chapter 3
Rainbow Schlieren

Keywords Rainbow filter · HSI scale · Abel transformation · Lorentz-Lorenz formula

3.1 Introduction

The laser schlieren arrangement described in Chap. 2 generates intensity contrast images using a knife-edge. The gradient normal to the knife-edge is highlighted in this approach. Gradients in other directions can be revealed by suitably orienting the knife-edge. This approach is not sufficiently general and one can look for other routes by which gradients can be imaged in one setting. Apart from this disadvantage, the knife-edge suffers from diffraction errors. These difficulties can be circumvented with the help of a graded filter wherein a transparent film with a grayscale imprinted on it is used at the knife-edge location. The film is a generalization of the knife-edge which can be understood as a *cut-off* filter with two shades, black and white. On a graded filter, the grayscale highlighted by the image can be used to determine the angle through which the light beam has been deflected. Extending this approach further, the gray-scale filter can be replaced by a *color filter* so that a color image is obtained from schlieren measurement. The implication here is that the laser is replaced by a white light source, and the monochrome CCD camera by a color CCD camera. When a color filter is used, the optical elements have to be achromatic. This approach is referred to as *rainbow* schlieren, and also referred as *color* schlieren since it generates a color image.

P. K. Panigrahi and K. Muralidhar, *Schlieren and Shadowgraph Methods in Heat and Mass Transfer*, SpringerBriefs in Thermal Engineering and Applied Science, DOI: 10.1007/978-1-4614-4535-7_3, © The Author(s) 2012

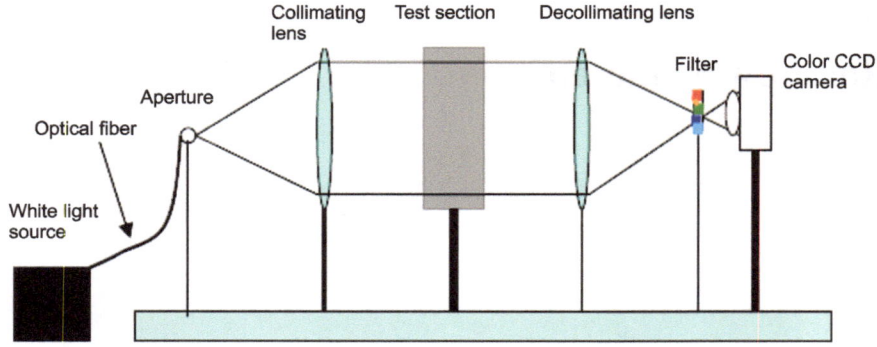

Fig. 3.1   Schematic diagram of a lens-type color schlieren setup

3.2 Optical Layout

A schematic drawing of the color schlieren apparatus is shown in Fig. 3.1.

The important elements of the set-up are white light source, collimating lens, decollimating lens, color filter, and color CCD camera. The optical components are mounted on a breadboard. The light source, camera, and the optical elements are aligned to have a common axis. The arrangement in Fig. 3.1 uses convex lenses to collimate and decollimate the light beam. For the applications discussed in later chapters, the lenses are respectively of 500 and 750 mm focal length, 65 and 100 mm diameter. The first (smaller focal length) lens acts as the collimator for light emerging from a pinhole, 100–200 μm diameter. While the second convex lens placed after the test cell decollimates the light beam. The color filter is placed at the focus of the decollimating lens. The test section is where the transport process is in progress. Light is deflected from its original path, and is re-directed by the second lens to form an image on the color filter. The initial image is the color distribution of the filter, in the absence of any density disturbance. The mount holding the color filter has an arrangement for both horizontal and vertical movements and these degrees of freedom are used during calibration. The color filter can also be moved around initially to select the initial color background in the image. If a 1-D color filter is used, it is positioned in such a way that the resulting color/hue variation is in the direction of significant density variation. The CCD camera placed beyond the filter captures images formed on the filter plane.

A 150 W, continuous, unpolarized, cold white light has been used as light source in the present work. The light source is connected to the pinhole using an optical fiber cable. While the pinhole approximates a point source, the lower limit of its diameter is fixed by the amount of light it can transmit for measurement. A 3-color CCD camera (*Basler*, A201bc) 648 × 648 pixels resolution was used in the present study. The CCD camera is connected to a personal computer through an 8/12-bit A/D card frame grabber. For the applications studied, a frame rate of 30 images per second was found to be adequate.

Table 3.1 R,G,B components of the VIBGYOR color distribution; from http://cloford.com/resources/colors/500col.htm

Color	Argument		
	Red	Green	Blue
Violet	238	130	238
Indigo	75	0	130
Blue	0	0	255
Green	0	255	0
Yellow	255	255	0
Orange	255	165	0
Red	255	0	0

As in all optical methods, the quality and content of color schlieren images depend on the quality of optical elements used and the degree of alignment. Issues such as quality of the light source, parallelism of the light beam, and diffraction rings formed at the pinhole have to be adequately addressed. Under ideal conditions, the initial (undisturbed) spot of light falling on the color filter should be identical in size and spectral content to the one leaving the pinhole. Once these conditions are fulfilled, the resolution of schlieren measurement is fixed by the design of the color filter.

3.3 Filter Design

In a 1D filter, the color variation is in one direction, say the vertical, while being a constant in the other. This design is most suitable for imaging flow fields in which gradients of concentration and temperature are mostly in one direction. In 2D filters, color varies in two orthogonal directions and two components of the property gradient can be obtained. The color images are recorded using a 3-color CCD camera and images are processed in terms of the hue distribution over the image plane.

3.3.1 1D Rainbow Filter

The design presented here follows the recommendation of [3]. The rainbow filter has a VIBGYOR variation of color with the resulting hue varying linearly in the direction of interest. Since the camera detects colors R (red), green (G), and blue (B), a connection between RGB and VIBGYOR is required. This relationship for 8-bit digitization (over 0-255 for each color) is given in Table 3.1. The R,G,B values of VIBGYOR are linearly interpolated to form an array of 1,200 × 1,200 entries. The color array of R,G,B values is used to generate a one-dimensional Cartesian filter. Figure 3.2a shows the photograph of a one-dimensional color filter developed

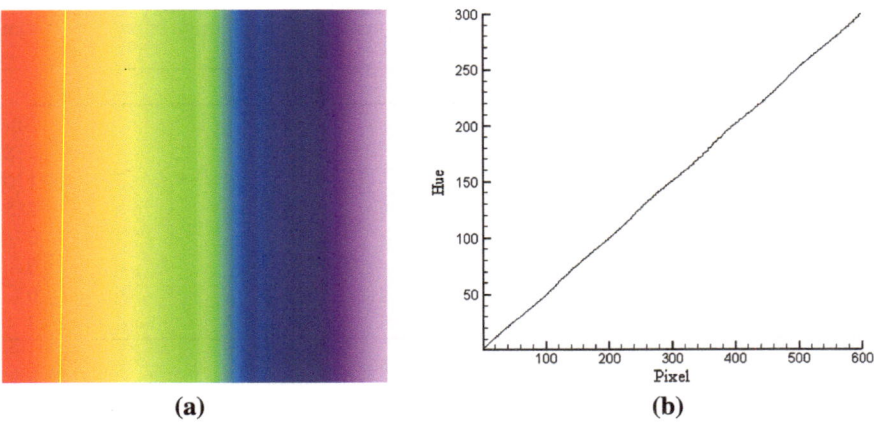

Fig. 3.2 **a** An image of 1D color filter and **b** Hue variation with the x coordinate (*in pixels*)

using this procedure. Here, MATLAB can be used for designing the filter and for image processing. It is now required to convert the color distribution available as data into a physical filter. An 8-bit image of the color filter is displayed on a computer screen and then recorded on a positive film (*Kodak Ektachrome*) in a dark room. The film, developed and placed in a 35 mm slide mount, is ready to work as a 1D color filter. Filters of varying sizes can be obtained when the computer screen is photographed from varying distances or lenses of varying focal lengths are used. 1D filters produced in this manner are quite inexpensive; a variety of filters can now be made and tailored to the measurement sensitivity required.

As discussed later in the section on data analysis, the color image is often analyzed in terms of hue (H) which in turn is related to beam deflection on the filter plane. Hue is measured as an angle in radians ($0–2\pi$) or degrees ($0–360°$). Figure 3.2b shows a plot of hue variation with position. A linear relation between the two can greatly simplify analysis. Ideally, the filter should possess the following property:

$$\frac{\partial H}{\partial x} = -\frac{2\pi}{X} = \text{constant}. \tag{3.1}$$

Here, x is the coordinate along which there is a change in color and X is the overall dimension of the filter. Departure from linearity can affect sensitivity and introduce distortion in data. Numerical methods can be used in such situations to rearrange colors and produce a linear hue variation.

3.3.2 2D Rainbow Filter

For a 2D color filter, color varies in two orthogonal directions. It is possible to measure light deflection with respect to the coordinates x and y and hence estimate

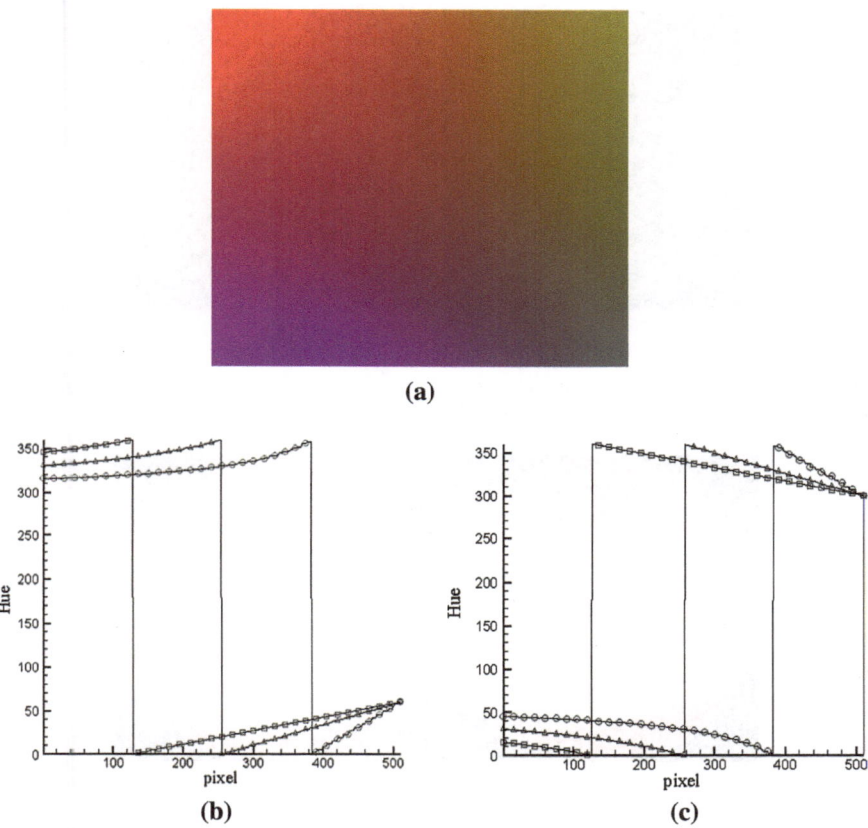

(a)

(b)   (c)

Fig. 3.3 **a** An example of 2D color filter; **b** Hue variation in the horizontal direction; and **c** Hue variation in the vertical direction.

gradients of the transported variable in two directions. The graded filter is designed to have colors change gradually over a square area. A relationship used in this work is of the form [2]:

$$R = \frac{255}{(1 + x_f + y_f)} \qquad G = \frac{255 y_f}{(1 + x_f + y_f)} \qquad B = \frac{255 x_f}{(1 + x_f + y_f)}. \qquad (3.2)$$

Equation 3.2 assumes 8-bit digitization of each of the colors. Here, x_f and y_f are scaled coordinates of the point at which R,G, and B values are calculated and fall in the range of 0–1. The construction of an image on the screen followed by its record on a photographic film proceeds as outlined for a 1D filter (Fig. 3.3).

Figures 3.4 and 3.5 show the baseline image of a 1D color filter with the colors running in the horizontal and vertical directions respectively. The image of a candle flame for each filter orientation is also seen. The image of the flame with a filter that

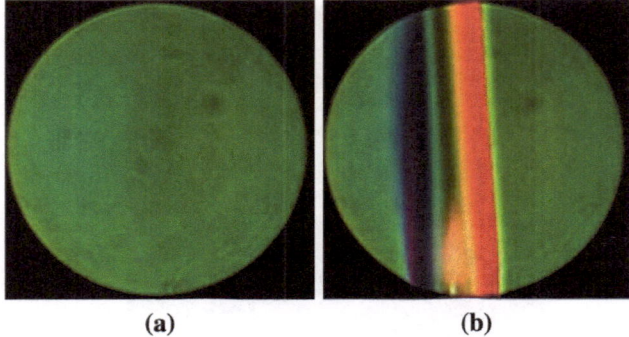

Fig. 3.4 a Base image recorded in 1D color schlieren when the filter has color variation in the vertical direction; **b** schlieren image of a candle flame

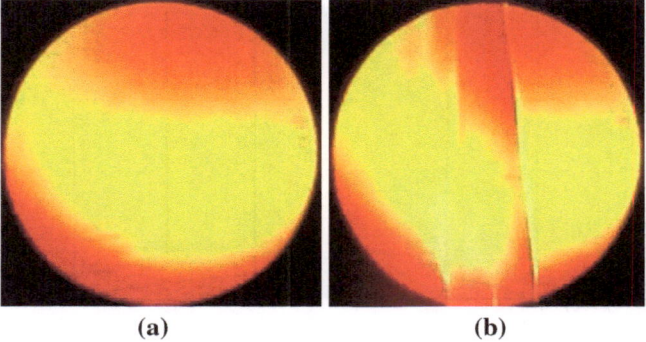

Fig. 3.5 a Base image recorded in 1D color schlieren when the filter has color variation in the horizontal direction; **b** schlieren image of a candle flame

has horizontal color strips (Fig. 3.4) shows vivid color patterns. This trend is expected because density gradients within the flame are strong in the vertical direction and the light beam is deflected across a range of colors. In Fig. 3.5, the light beam stays within a color band and the image is closer to that seen in monochrome schlieren.

3.4 HSI Parameters

The human visual system can distinguish thousands of varying color shades and intensities but only around 100 shades of gray. Therefore, in an image, a great deal of extra information is contained in the form of color. The extra information can be used to simplify image analysis such as object identification and extraction. Color sensation in a color schlieren measurement is generated by a number of factors. These include the design (and sensitivity) of the color filter, the spectral characteristics of light used for illumination, and the spectral response of the color-sensing detector which could be a human eye or an imaging sensor in a color camera. A color model is simply a convenient way to represent color in numerical terms. Most color models use a 3D coordinate system. Each point within the system's subspace represents a unique color. The RGB color model, for example, can be visualized as a cube where red is the x-axis, blue is the y-axis, and green is the z-axis. Each one of the many million colors possible is described as a unique point within the cube.

There are many other color models in use. Apart from RGB, other possibilities include the HSI model (Hue, Saturation, Intensity) and the HSV color model (Hue, Saturation, Value). These are frequently used in digital image processing. The RGB model is difficult to use directly because it requires three values of red, green, and blue to interpret density gradient at a pixel. It is preferable to have a quantity that will have a single value corresponding to a particular density gradient. The HSI model serves this purpose as discussed next. In the HSI model, color is specified by the three quantities, namely hue, saturation, and intensity. In the visible spectrum, hue directly corresponds to the dominant wavelength of color. Saturation refers to the degree to which a color deviates from a neutral gray of equal intensity. Saturation may also be defined as color's purity or the amount of white contained in a specific color. When highly desaturated, any color of the spectrum should approach the standard white color. The analogy here is to white noise when the signal strength is progressively reduced. Intensity of a color refers to its relative brightness in the color mixture. It represents spectral energy at the specific wavelength arriving at the sensor. The HSI model can be represented in terms of the color-space by defining a 3D cylindrical coordinate system. The hue distribution is represented as an angle varying from 0 to 360°. Saturation corresponds to the radius, varying from 0 to 1. Intensity varies along the z-axis with 0-black to 1-white. Adjusting hue will vary color from red at 0 through green at 120, blue at 240, and back to red at 360°.

The advantage of using the HSI model is that hue is insensitive to absolute light intensity. This property eliminates numerous complications such as intensity fluctuations of the source, pixel-to-pixel variation in gain within a given detector array or from one detector to another, absorption or scattering of light within the test section, or second-order intensity variations resulting from asymmetric refractive index distribution. In addition, hue scales strongly (and uniquely) with beam deflection and simplifies data analysis. The HSI parameters can be obtained directly from the RGB values recorded by the camera from the following Eq. [3]:

$$I = \frac{R + G + B}{3}, \tag{3.3}$$

$$S = 1 - \frac{\min(R, G, B)}{I}, \tag{3.4}$$

$$H = \cos^{-1}\left(\frac{\frac{1}{2}[(R - G) + (R - B)]}{[(R - G)^2(R - G)(G - B)]^{\frac{1}{2}}}\right). \tag{3.5}$$

Fig. 3.6 Graphical representation of the HSI color space
and its relationship to the R,
G, and B tristimus vectors
(after [3])

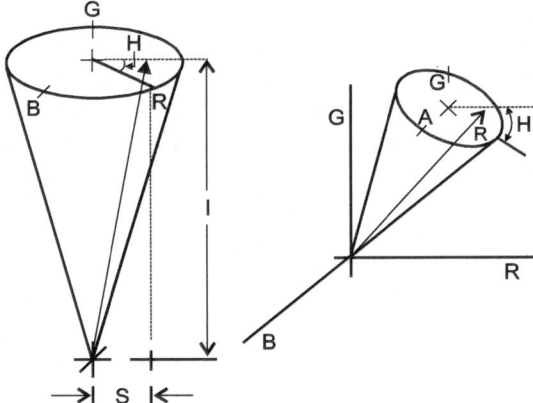

A schematic representation of the relationship between the RGB and HSI color spaces
is shown in Fig. 3.6.

The vertex of the HSI cone is anchored at the intersection of the R, G, B axes, and
the projections of the central axis of the cone onto the R, G, and B axes are identical.
Equation 3.5 shows that hue is independent of absolute intensity. This is also seen
in Fig. 3.6 where, hue, given as the polar angle relative to pure red, is unaffected
by the total length (intensity) of the color-space vector. Thus the RGB output of a
conventional color imaging array can be readily transformed into a suitable one-
parameter measure of color. Thus,on one of the lateral dimensions of the filter plane,
a band-like filter is produced. Similar to knife-edge, such a filter is sensitive to only
the horizontal (or vertical) component of deflection. When mapped onto the radial
coordinate, an axisymmetric filter is produced. A filter of this type responds to the
absolute magnitude of the resulting deflections, but not to the orientation as hue will
be a constant at every angle for a given radius.

3.4.1 Calibration of 1D Filter

The calibration curve of a 1D color filter is a relationship between hue and light beam
displacement. It is obtained from the experimental apparatus itself, with thermal
disturbances absent in the test cell. Initially the filter is placed at the focus of the
decollimating lens. The 1D color filter can be traversed in the vertical direction if the
expected beam displacements are mainly along the vertical axis. For every position,
the camera records the RGB image of the filter. The filter position is monitored by a
micrometer arrangement. The color data is then converted into HSI coordinates. The
calibration curve is the relationship between hue and the micrometer reading. The
position of the spot of light that falls on the axis of the apparatus can be taken as the
datum since it corresponds to zero density gradients in the experiment.

3.5 Image Formation in Color Schlieren

This section describes the process of image formation in a color schlieren measurement. With this formulation, density gradients (equivalently, temperature and concentration gradients) can be determined from the hue distribution in the image plane. As discussed in Chap. 1, refractive index techniques are based on the Lorentz-Lorenz formula

$$\frac{1}{\rho}\frac{(n^2-1)}{(n^2+2)} = \text{constant}, \tag{3.6}$$

where n is refractive index and ρ, the density. For gases the refractive index is close to unity, i.e., $n \cong 1$ and Eq. 3.6 reduces to the Gladstone-Dale relation

$$\frac{n-1}{\rho} = G. \tag{3.7}$$

The constant G, called Gladstone-Dale constant is a function of the chemical composition of the gas. It varies slightly with the wavelength of light. Instead of using G directly, one can use the reference condition marked "0" and write

$$n - 1 = \frac{\rho}{\rho_0}(n_0 - 1) \tag{3.8}$$

or

$$\rho = \rho_0 \frac{n-1}{n_0 - 1}. \tag{3.9}$$

It is clear that $dn/d\rho$ is constant for gases.

In schlieren, the first derivative (say, with respect to coordinate y) is determined and Eq. 3.7 is written as

$$\frac{\partial \rho}{\partial y} = \frac{1}{G}\frac{\partial n}{\partial y} = \frac{\rho_0}{n_0 - 1}\frac{\partial n}{\partial y}. \tag{3.10}$$

To make temperature measurement possible, the refractive index variation must be related to that of temperature. For moderate changes in temperature, typically $\leq 20K$ and nearly uniform bulk pressure, gas density varies linearly with temperature as

$$\rho = \rho_0(1 - \beta(T - T_0)). \tag{3.11}$$

Accordingly, refractive index varies linearly with temperature as well; see also Sects. 1.4.1 and 2.2.2.

For a process involving mass transfer, Lorentz-Lorenz formula as applied to a solute–solvent system takes the form [6]:

$$\frac{n^2 - 1}{n^2 + 2} = \frac{4}{3}\pi(\alpha_A C_A + \alpha_B C_B). \tag{3.12}$$

Here, n is the refractive index of the solution, and α and C are respectively the polarizability and the number of moles of salt in the solution. In the crystal growth application (Chap. 8), suffixes A and B specify water as the solvent and KDP as the solute, respectively [4]. The corresponding relationship between the gradients of salt concentration and refractive index is given by:

$$\frac{\partial C}{\partial y} = \frac{9n}{2\alpha_{KDP}(n^2 + 2)^2}\frac{\partial n}{\partial y}. \tag{3.13}$$

Here, α_{KDP} is the polarizability of the KDP crystal ($=4.0\,\mathrm{cm}^3/\mathrm{mole}$) and C is the molar concentration of the solution (moles per 100 gm of the solution). Equation 3.13 can be integrated from a location in the bulk of the solution (where concentration gradients are negligible). Concentration distribution around the growing crystal can now be uniquely determined.

Changes in refractive index in the fluid region are determined in terms of beam deflection, and hence changes in hue in the color filter. The relationship between the two is developed as follows. Let the z-axis be the direction of propagation of the undisturbed ray, normal to the x-y plane. The cumulative angular deflections measured beyond the test section, denoted respectively as α_x and α_y can be derived as in Chap. 2 to yield:

$$\alpha_x = \frac{1}{n_0}\int\frac{\partial n}{\partial x}\mathrm{d}z \tag{3.14}$$

$$\alpha_y = \frac{1}{n_0}\int\frac{\partial n}{\partial y}\mathrm{d}z, \tag{3.15}$$

where n_0 is the refractive index of air surrounding the test section. The light beam turns in the direction of increasing index of refraction. In most media this means that the light is bent towards the region of higher density (Fig. 3.7).

Light passing through each section of the test region comes from all parts of the source. Thus, at the focus not only is the image of the source composed of light coming from the whole field of view, but light passing through every point in the field of view gives an image of the source at the filter plane. If light from a position x, y in the test region is deflected by an angle α, then the image of the source coming from that position will be shifted at the filter plane by an amount

$$\Delta a = \pm f_2 \tan\alpha \approx \pm f_2\alpha, \tag{3.16}$$

where the sign is determined by the change in hue at the filter plane. In color schlieren, the deflection of the light beam effectively leads to a change in hue. If the light beam at any position x, y in the test section is deflected by an angle α, then the shift in the light beam at the color filter is given by

$$\Delta a_x = f_2\alpha_x, \tag{3.17}$$

Fig. 3.7 Undeflected and deflected light beams shown at the color filter of a color (*rainbow*) schlieren system

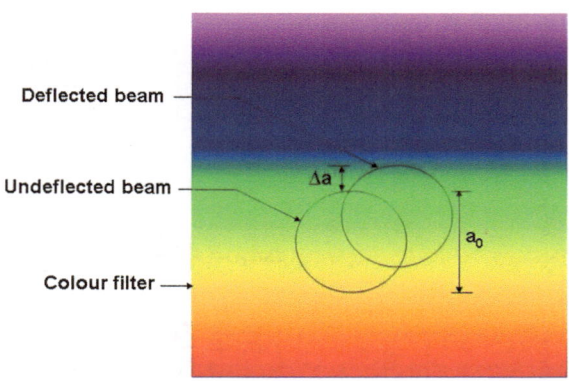

Deflected beam

Undeflected beam

Colour filter

$$\Delta a_y = f_2 \alpha_y. \tag{3.18}$$

Quantities Δa_x and Δa_y can be obtained from the variation of hue at the x, y position of the image via the calibration curve. A 2D filter would be required to capture both components. Thus, a difference in hue will give the absolute deflection in each of the directions. From Eqs. 3.14–3.18

$$\Delta a_x = \frac{f_2}{n_0} \int_0^L \frac{\partial n}{\partial x} dz, \tag{3.19}$$

$$\Delta a_y = \frac{f_2}{n_0} \int_0^L \frac{\partial n}{\partial y} dz. \tag{3.20}$$

Assuming a 2D field of length L in which $n = n(x, y)$ with no dependence on z, the viewing direction, we get

$$\Delta a_x = \frac{f_2}{n_0} \frac{\partial n}{\partial x} L, \tag{3.21}$$

$$\Delta a_y = \frac{f_2}{n_0} \frac{\partial n}{\partial y} L. \tag{3.22}$$

Using Eq. 3.10, the above equation can be rewritten for gaseous media as

$$\Delta a_x = \frac{f_2}{n_0} \frac{n_0 - 1}{\rho_0} \frac{\partial \rho}{\partial x} L, \tag{3.23}$$

$$\Delta a_y = \frac{f_2}{n_0} \frac{n_0 - 1}{\rho_0} \frac{\partial \rho}{\partial y} L. \tag{3.24}$$

For ideal gas at constant pressure (P), these equations can be expressed as

Fig. 3.8 Representation of an
axisymmetric refractive index
field, after [1]

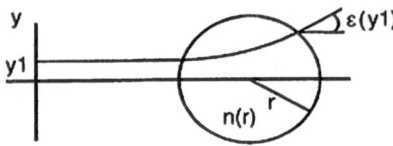

$$\Delta a_x = \frac{f_2}{n_0} \frac{n_0 - 1}{\rho_0} \frac{P}{RT^2} \frac{\partial T}{\partial x} L, \tag{3.25}$$

$$\Delta a_y = \frac{f_2}{n_0} \frac{n_0 - 1}{\rho_0} \frac{P}{RT^2} \frac{\partial T}{\partial y} L. \tag{3.26}$$

The beam deflection in a KDP solution is obtained by combining Eqs. 3.13 and 3.20
as:

$$\Delta a_y = \frac{f_2}{n_0} L \frac{2\alpha_{KDP}(n^2 + 2)^2}{9n} \frac{\partial C}{\partial y}. \tag{3.27}$$

Along with a boundary condition, Eq. 3.27 can be integrated from a location in the
bulk of the solution (where the gradients are negligible) to derive the concentration
distribution around the crystal.

The material property that determines the sensitivity of the optical measurement
is dn/dC or dn/dT, where n is the refractive index, C is the solute concentration,
and T is temperature. Compared to gases (e.g. air), its value is about three orders
of magnitude larger in liquids. Hence, only a small perturbation is needed to see
refraction in liquids compared to air.

3.5.1 Analysis of an Axisymmetric Field

In several applications, the physical domain is circular and the quantity of interest
has an axisymmetric distribution with respect to the radial coordinate. An example is
the initial development of a gas jet in air, where the property of interest is gas density,
equivalently, gas concentration with respect to air [8]. A model schlieren arrangement
for such a configuration is shown in Fig. 3.8, where r is the radial coordinate. The
field is 2D in Cartesian coordinates but 1D in polar coordinates. The refractive index
field $n(r)$ within the circular region is to be determined from the schlieren image.
A light beam starting at a location $y1$ along the y-axis emerges at an angle $\epsilon(y1)$ at
the boundary of the physical domain.

For the axisymmetric refractive index field shown in Fig. 3.8, it can be shown
that [7]

$$\epsilon(y) = 2y \int_y^\infty \frac{d\delta}{dr} \frac{dr}{(r^2 - y^2)^{0.5}}. \tag{3.28}$$

Here, $\delta = \eta - 1$ is the refractive index difference relative to vacuum and η is the refractive index of the test medium normalized by that of the surrounding air. It is given as

$$\eta = \frac{n_{\text{medium}}}{n_{\text{ambient}}}.$$

A commonly used value is $n_{\text{ambient}} = 1.0003$. With f_c, the focal length of the decollimating lens, the transverse displacement of a ray of light at the filter plane is given by

$$\delta(y) = f_c \epsilon(y). \tag{3.29}$$

The light deflection data $\delta(y)$ can be obtained from the hue distribution recorded from a 1D color filter whose color varies in the y-direction. From Eq. 3.29, the deflection angle $\epsilon(y)$ can thus be recovered. The refractive index field is now obtained by inverting Eq. 3.28 using Abel transformation [6] as follows:

$$\delta(r) = -\frac{1}{\pi} \int_r^\infty \epsilon(y) \frac{dy}{(y^2 - r^2)^{0.5}}. \tag{3.30}$$

The integral of Eq. 3.30 is split as a sum of integrals by factoring out the deflection angle [9]. Thus,

$$\delta(r_i) = -\frac{1}{2\pi} (\epsilon_j + \epsilon_{j+1}) \int_{r_j}^{r_j+1} \frac{dy}{(y^2 - r_i^2)^{0.5}}. \tag{3.31}$$

Here, $r_i = i \Delta r$ is the radial distance of the point of interest from the centerline and Δr is the sampling interval. The total number of intervals in the test region is indicated as N. Equation 3.31 can be presented in the form of a numerical algorithm for the refractive index difference δ as follows:

$$\text{Evaluate} \quad \delta(r_i) = \sum_i^N D_{ij} \epsilon_j \tag{3.32}$$

$$\begin{aligned} D_{ij} &= J_{ij} & \text{if } j = 1 \\ D_{ij} &= J_{ij} + J_{ij+1} & j > i \end{aligned} \tag{3.33}$$

with

$$J_{ij} = -\frac{1}{2\pi} \ln \left[\frac{(j+1) + [(j+1)^2 - i^2]^{0.5}}{j + (j^2 + i^2)} \right]. \tag{3.34}$$

It may be noted that the quantities D_{ij} are independent of Δr, the sampling interval.

In the context of buoyant jets [1, 8], the refractive index of a mixture of gases is given as [10]:

Table 3.2 Gladestone Dale constants for various gases, from [5]

Gas	Gladstone dale constant, $k(m^3/kg)$
Oxygen	0.190×10^{-3}
Nitrogen	0.238×10^{-3}
Hydrogen	1.5×10^{-3}
Helium	0.196×10^{-3}

$$n = 1 + \delta = 1 + \sum_i k_i \rho_i. \tag{3.35}$$

In the above expression the summation is over all the species present in the flow field. Symbols k_i and ρ_i represent Gladstone-Dale constant (Table 3.2) and density of the individual species respectively.

For ideal gases mixing under atmospheric conditions, Eq. 3.35 reduces to the form [10]

$$n = 1 + \frac{P}{RT} \sum k_i x_i M_i \tag{3.36}$$

Here, P is atmospheric pressure, R is universal gas constant, x_i is the mole fraction of the ith species, and M_i, the molecular weight.

3.6 Color Versus Monochrome Schlieren

There are several potential advantages of color schlieren over the monochrome schlieren technique. It replaces knife-edge with a color filter. It avoids diffraction effects present in a monochrome schlieren image at the knife-edge. Diffraction errors may necessitate complex corrections in monochrome schlieren. The difficulty with camera saturation in monochrome schlieren arises from the use of a laser and is greatly reduced in color schlieren due to a white light source. A color filter measures positive as well as negative displacements and permits extraction of quantitative information such as gradient magnitudes and their direction. Color contrast can discriminate minute features and distinguish authentic beam displacement from the silhouettes of opaque or spurious objects in the field of view. Extending the measurement to a 2D color filter, x- and y- gradient information can be recorded. These advantages have to weighed against the need to design and calibrate a color filter of suitable resolution and the cost of a color CCD camera. For an assessment of color schlieren against BOS, see [2].

References

1. Al-Ammar K, Agrawal AK, Gollahalli SR, Griffin D (1998) Application of rainbow schlieren deflectometry for concentration measurements in an axisymmetric helium jet. Exp Fluids 25:89–95
2. Elsinga GE, Oudheusden BW, Scarno VF, Watt DW (2003) Assessment and application of quantitative schlieren methods with bi-directional sensitivity: CCS and BOS, In: Proceedings of PSFVIP-4, Chamonix, France, pp 1–17
3. Greenberg PS, Klimek RB, Buchele R (1995) Quantitative rainbow schlieren deflectometry. Appl Opt 34(19):3870–3822
4. Gupta AS (2011) Optical visualization and analysis of protein crystal growth process. Ph.D. dissertation, GB Technical University, India, submitted
5. Merzkirch W (1974) Flow visualization. Academic Press, New York and London
6. Mantani M, Sugiyama M, Ogawa T (1991) Electronic measurement of concentration gradient around a crystal growing from a solution by using Mach-Zehnder interferometer. J Crystal Growth 114:71–76
7. Rubinstein R, Greenberg PS (1994) Rapid inversion of angular defection data for certain axisymmetric refractive index distributions. Appl Opt 33:1141–1144
8. Semwal K (2008) Jet mixing study using color schlieren technique: influence of buoyancy and perforation. Master's Thesis, IIT Kanpur (India)
9. Vasil'ev LA (1971) Schlieren methods. Israel program for scientific translation, pp 176–177, Springer, New York
10. Yates LA (1993) Constructed interferograms, schlieren and shadowgraphs: a user's manual, NASA CR-194530

Chapter 4
Principles of Tomography

Keywords CBP · ART · MART · Entropy · Sensitivity · Extrapolation scheme

4.1 Introduction

Most physical systems involving heat and mass transfer have 3D variation of temperature and species concentration within the apparatus. When schlieren or shadowgraph imaging is employed, one obtains a depth-averaged view of the 3D variation. The image data is often called a path integral or a projection of the thermal (or concentration) field. Tomography is a procedure for recovering the 3D information of the field variable from a collection of projections. The projection data is recorded at various angles by turning the experimental apparatus or the light beam, specifically the source-detector axis. This chapter presents tomography in the form of algorithms, of which CBP, ART, and MART are a few. Tomographic algorithms are known to be sensitive to noise in projection data. Issues such as sensitivity and the impact of having limited data are discussed. The algorithms are validated using simulated data as well as from physical experiments of crystal growth and jet interactions. Finally, the use of POD as a tool for dealing with unsteady data is presented.

4.2 Overview

Refractive index based techniques discussed in the present monograph generate path integrals of the 3D refractive index field (and hence density) in the direction of propagation of light through the apparatus. Equivalently, one obtains path integrals of the 3D temperature or concentration field. The path integrals are also referred to as *projection data*. These integrals are seen as 2D images over a screen. When

P. K. Panigrahi and K. Muralidhar, *Schlieren and Shadowgraph Methods in Heat and Mass Transfer*, SpringerBriefs in Thermal Engineering and Applied Science, DOI: 10.1007/978-1-4614-4535-7_4, © The Author(s) 2012

temperature and concentration are jointly present, a dual wavelength laser can be used to distinguish between their respective signatures. The question addressed here is the recovery of 3D information from 2D images.

When combined with holography, laser interferometry can be extended to map 3D fields [12]. However, holographic interferometry can be cumbersome in some applications due to the need of holographic plates, particularly when large regions have to be scanned. As discussed in the present chapter, this difficulty is circumvented by using an analytical technique called tomography. Furthermore, a holographic approach to schlieren and shadowgraph is not available while tomography continues to be applicable.

In a schlieren image, the projection data is in the form of a refractive index gradient and hence gradient in temperature and concentration, in a direction fixed by the knife-edge or the gray scale/color filter. The gradient information can be integrated to yield data in terms of temperature and concentration. Several such projections can be recorded at various angles by turning the apparatus (the preferred route) or the source-detector axis. From the images recorded, it is possible to recover the entire 3D field using principles of *tomography*. The process of recording projections involves passage of time and the overall process is useful only in steady contexts. The extension of tomography to unsteady problems is a topic of research. A possible route to dealing with unsteadiness is discussed toward the end of this chapter.

In optical techniques discussed in the present monograph, the light source (laser) and the detector (CCD camera) lie on a straight line with the test cell in between while a parallel beam of light is used. This configuration is called *transmission* tomography and the ray configuration is a *parallel beam geometry*. Other arrangements such as fan-beam and cone-beam tomography are encountered in medical applications and are not within the scope of the present discussion.

Tomography is the process of recovery of a function from a set of its line integrals evaluated along well-defined directions. It can be classified into: (a) transform, (b) series expansion, and (c) optimization methods. Transform methods are direct, non-iterative, and do not need an initial guess. As a rule, they require a large number of projections for a meaningful retrieval of the 3D field. In practice, projections can be recorded either by turning the experimental setup or the source-detector combination. With the first option, it is not possible to record a large number of projections, partly owing to inconvenience and partly to time and cost. Hence, as a rule, a large number of projections cannot be acquired and one must look for methods that converge with just a few projections. Limited-view tomography is best accomplished using the series expansion method. As limited-view tomography does not have a unique solution, the algorithms are expected to be sensitive to the initial guess of the field that start the iterations. Optimization-based algorithms are known to be independent of initial guess, but the choice of the optimization function plays an important role in the result obtained. Depending on the mathematical definition used, the entropy extremization route may yield good results, while the energy minimization principle may be suitable in other applications.

For transform techniques, an unbiased initial guess such as a constant profile may be good enough for accurate reconstruction. A complete random number guess can

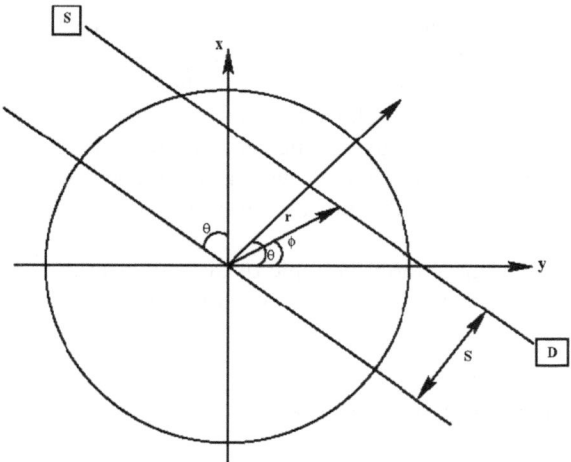

Fig. 4.1 Schematic drawing of data collection using the parallel beam geometry. S-source, D-detector, s-perpendicular distance from the center of the object to the ray, θ-view angle, and (r, ϕ)-polar coordinates

also be viewed as an unbiased initial guess. Tomography is an inverse technique and is sensitive to errors in the projection data. However, it can reveal dominant trends in the field variable after inversion.

In laser schlieren setup, the projection data of the 3D refractive index field is recorded in the following manner. The source of light (the laser) and the detector (the CCD camera) lie on a straight line with the test chamber in-between. A parallel beam of light is used to scan the field of interest. The recording configuration is shown in Fig. 4.1. The tomography algorithms reconstruct 2D fields from their 1D projections. Reconstruction can then be sequentially applied from one plane to the next until the third dimension is filled. Tomography is the process of recovery of a function from a set of line integrals, namely projections evaluated along well-defined directions. A projection of a 2D function is a line integral in a certain direction. Specifically, let a given object be prescribed as the function $f(r, \phi)$. For a single data ray SD propagating along the z-direction, the projection data $p(s, \theta)$ is given as

$$p(s, \theta) = \int_{SD} f(r, \phi) \mathrm{d}z. \tag{4.1}$$

The ray indices are s and θ, where s is the perpendicular distance of the ray from the object center and θ is the angle of the source position (or object rotation). Here, z is a coordinate along the chord SD. In a given experiment, optical techniques collect the projection data $p(s, \theta)$ for various values of s and for several θ in the form of images. The projection data is recorded for θ ranging from 0 to π. The transformation of an object function $f(r, \phi)$ into its projection data $p(s, \theta)$ is called the *Radon Transform*. Equation 4.1 is the Radon transform of the object function $f(r, \phi)$. Tomography is concerned with recovering function $f((r, \phi)$ from the data $p(s, \theta)$ [10, 14, 15]. The reconstruction then fills the r-ϕ or x-y plane. The process is repeated for successive planes above to create a truly 3D distribution of the field variable. The utility of tomography in optical method stems from the fact that optical images such as interferograms, schlieren, and shadowgraph relate to the field variable via path integrals evaluated in the direction of propagation of light, as discussed in Chaps. 1–3.

Tomographic algorithms form the topic of discussion in the following section. The validation of these techniques is subsequently presented.

4.3 Convolution Backprojection

The convolution backprojection (CBP) algorithm for 3D reconstruction of an object from its projections classifies as a transform technique. It has been used for medical imaging of the human brain for the past several decades. Significant advantages of this method include (a) its non-iterative character, (b) availability of analytical results on convergence of the solution with respect to the projection data, and (c) established error estimates. A disadvantage to be noted is the large number of projections normally required for good accuracy. In engineering applications, this translates to costly experimentation, and non-viability of recording data in unsteady experiments. The use of CBP continues to be seen in steady flow experiments, particularly when the region to be mapped is physically small in size. The statement of the CBP algorithm is presented below.

Let the path integral Eq. 4.1 be re-written as

$$p(s, \theta) = \int_C f(r, \phi) \mathrm{d}z, \qquad (4.2)$$

where p is the projection data recorded in the experiments and f is the function to be computed by inverting the above equation. In practice, the function f is a field variable, such as density, void fraction, attenuation coefficient, refractive index, temperature, or concentration. The symbols s, θ, r, and ϕ stand for the ray position, view angle, position within the object to be reconstructed, and the polar angle, respectively (Fig. 4.1). Integration is performed with respect to the variable z along the chord C of the ray defined by s and θ. Following Herman [10], the *projection slice theorem* can be employed in the form

$$\overline{p}(R, \theta) = \overline{f}(R\cos\theta, R\sin\theta), \qquad (4.3)$$

where the overbar indicates the Fourier transform and R is the spatial frequency. In words, this theorem states the equivalence of the 1D Fourier transform of $p(s, \theta)$

with respect to s and the 2D Fourier transform of $f(r, \phi)$ with respect to r and ϕ. A 2D Fourier inversion of this theorem leads to the well-known result [10]

$$f(r, \phi) = \int_0^\pi \int_{-\infty}^\infty \overline{p}(R, \theta) \exp(i2\pi Rr \cos(\theta - \phi)) |R| dR d\theta$$

where $i = \sqrt{-1}$ and

$$\overline{p}(R, \theta) = \int_{-\infty}^\infty p(s, \theta) \exp(-i2\pi Rs) ds.$$

The first integral in the form given above is divergent with respect to the spatial frequency R. Practical implementation of the formula replaces $|R|$ by $W(R)|R|$, where W is a window function that vanishes outside the interval $[-R_c, R_c]$. The cut-off frequency R_c can be shown to be inversely related to the ray spacing for a consistent numerical calculation of the integral. When the filter is purely of the band-pass type, the inversion formula can be cast as a convolution integral [10, 23]:

$$f(r, \phi) = \int_0^\pi \int_{-\infty}^\infty p(s, \theta) q(s' - s) ds d\theta, \qquad (4.4)$$

where

$$q(s) = \int_{-\infty}^\infty |R| W(R) \exp(i2\pi Rs) dR$$

and

$$s' = r \cos(\theta - \phi)$$

The inner integral over s is a 1D convolution and the outer integral, an averaging operation over θ is called *back projection*. This implementation of the convolution backprojection algorithm is commonly used in medical imaging.

Applications of the CBP algorithm to flow and heat transfer problems have been reviewed in [22].

4.4 Iterative Techniques

Series expansion methods are perhaps the most appropriate tomographic algorithms for refractive index-based techniques since they work with limited projection data

[2, 3, 5, 13, 16, 21, 24, 26–28, 32, 33, 36]. These methods are iterative in nature and consist necessarily of four major steps, namely:

1. Initial assumption of the field to be reconstructed over a grid,
2. Calculation of the correction for each pixel,
3. Application of the correction, and
4. Test for convergence.

The central idea behind the calculation of the correction (step 2) is the following. With the assumed field, one can explicitly compute the projection values by numerical integration. The difference between the computed projection and experimentally recorded projection data is a measure of the error in the assumed solution. This error can be redistributed to the pixels so that error is reduced to zero. Repetition of these steps is expected to converge to a meaningful solution. The series expansion techniques differ only in the manner in which the errors are redistributed over the grid.

The word *convergence* in step 4 is used in an engineering sense as a stopping criterion for the iterations, and not in the strict mathematical sense, where a formal proof is needed to show convergence of the numerical solution to the exact.

Iterative methods require the discretization of the plane to be reconstructed on a rectangular grid, Fig. 4.2. The length of the intercept of the ith ray with the jth cell in a given projection is known as the weight function, w_{ij}. If f_j is the field value in the jth cell, the path integral can be approximated as a sum and it can be shown that

$$\phi_i = \sum_{j=1}^{N} w_{ij} f_j \quad i = 1, 2, \ldots M, \tag{4.5}$$

where ϕ refers to the projection data. The number of unknowns N is, in general not equal to the number of rays M. This discretization produces a matrix equation

$$[w_{ij}]\{f_j\} = \{\phi_i\}. \tag{4.6}$$

The problem of reconstruction thus is a problem of inversion of a rectangular matrix [4]. Iterative techniques that are used in the tomography can be viewed as developing a generalized inverse of the matrix $[w_{ij}]$. This is a sparse matrix with many of its element being zero. General purpose matrix libraries cannot be used to invert such matrices since they are highly ill-conditioned and rectangular in structure. Tomographic algorithms can be seen as a systematic route toward a meaningful inversion of the matrix equation.

Series expansion methods being discussed in the present section can be classified into: ART (Algebraic Reconstruction Technique) and MART (Multiplicative Algebraic Reconstruction Technique). Optimization techniques that are also iterative in nature are discussed with reference to maximization of entropy and minimization of energy.

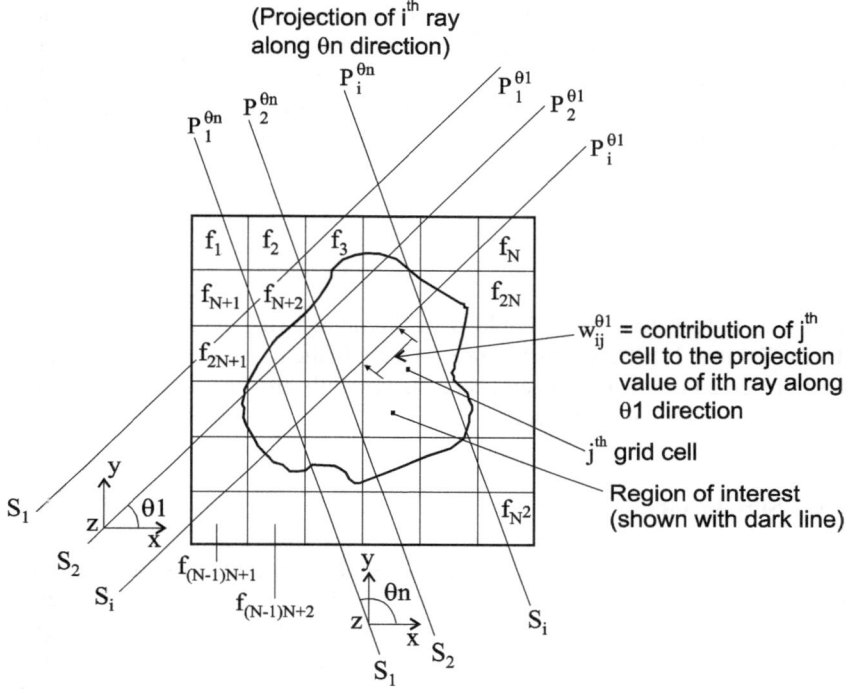

Fig. 4.2 Discretization of a plane of fluid with temperature or concentration distribution for iterative calculations. The viewing direction is indicated as θ, the cell index is j and the ray index is i

The ART and MART families of algorithms differ only in the method of updating the field parameters in each iteration. In ART, the correction is additive while for MART, the correction is multiplicative. In both cases, the numerical procedure is based on the comparison of the estimated projection from an initial guess with the measured projection data obtained through experiments. This gives a correction term for the field variables. The value of the field variables are then updated. Once an iteration is over, the field value differs from the previous guess. The extent of the difference is then calculated. If the difference is within acceptable limits, the field value is taken to represent the physical field. Otherwise iterations continue till the convergence criterion is satisfied.

Since the original field in real experiments is unknown, an estimate of the number of iterations can be found by using test functions (called *phantoms*) that are similar in nature to the original field. The test functions are also perturbed with noise to gauge the sensitivity of the algorithms to issues such as initial guess and errors in the projection data. This method can only be adopted where an exact estimate of noise in the projection data and a good knowledge of the original field is known before hand. Variations in the noise level and nature of distribution of the noise in the projection data can alter the convergence rates.

Tomographic algorithms referred in the present monograph are iterative in nature and intermediate steps may also involve iterations in the form of FOR loops. To identify the beginning and the ending of each iterative loop, start and close labels with statement numbers are indicated in the description of each algorithm. These algorithms, in the form of pseudo codes, are briefly surveyed in the following sections.

4.4.1 ART

Various ART algorithms are available in the literature owing their origin to Kaczmarz [11] and Tanabe [31]. They differ from each other in the way the correction is applied. Those presented below have been tested successfully by the authors in the context of optical measurement techniques.

4.4.1.1 Simple ART

Suggested by Mayinger [14], corrections are applied through a weight factor, computed as an average correction along a ray. The difference between the calculated projections with the measured projection data gives the total correction to be applied for a particular ray. The average correction is then the contribution to each cell falling in the path of the ray. This is computed by dividing the total correction obtained with the length of the ray. The calculated projections are computed once for a particular angle. Though the field values are continuously updated the calculated projection values remain unchanged till the completion of all the rays for a given angle. This algorithm is referred to as ART1 in following discussion.

Let $\phi_{i\theta}$ be the projection due to the ith ray in the θ direction and $\tilde{f}_i$ be the initial guess of the field value. Numerically the projection $\tilde{\phi}_{i\theta}$ using the current field values is defined as:

$$\tilde{\phi}_{i\theta} = \sum_{j=1}^{N} w_{i\theta,j} f_j \quad i\theta = 1, 2 \ldots, M_\theta. \tag{4.7}$$

The individual steps in the algorithm are listed below.
Calculate the total weight function $(W_{i\theta})$ along each ray as:
start: 1 For each projection angle (θ):
start: 2 For each ray $(i\theta)$:
start: 3 For each cell (j):

$$W_{i\theta} = \sum_{j=1}^{N} w_{i\theta,j}$$

close: 3
close: 2

close: 1
start: 4 Start iterations (k):
start: 5 For each projection angle (θ):
start: 6 For each ray ($i\theta$):
Compute the numerical projection (Eq. 4.7)
close: 6
start: 7 For each ray ($i\theta$):
Calculate the correction as:

$$\Delta\phi_{i\theta} = \phi_{i\theta} - \tilde{\phi}_{i\theta}.$$

Calculate the average value of correction as:

$$\overline{\Delta\phi_{i\theta}} = \frac{\Delta\phi_{i\theta}}{W_{i\theta}}$$

close: 7
start: 8 For each ray ($i\theta$):
start: 9 For each cell (j):
If $w_{i\theta,j}$ is nonzero then:

$$f_j^{\text{new}} = f_j^{\text{old}} + \mu\overline{\Delta\phi_{i\theta}},$$

where μ is a relaxation factor.
close: 9
close: 8
close: 5
With ϵ as the prescribed convergence criterion, say 0.01 %, check for convergence
as:

$$\text{abs}\left[\frac{f^{k+1} - f^k}{f^{k+1}}\right] \times 100 \leq e \tag{4.8}$$

STOP:
Else: Continue
close: 4(k)

4.4.1.2 Gordon ART

The ART algorithm contributed by Gordon et al. [7, 8] is now considered. Mayinger's
ART is similar to this original version under the condition that no two rays simultaneously pass through a particular cell for a given projection. In this method corrections
are applied to all the cells through which the ith ray passes, using the weight factor
which is exactly the proportion of w_{ij} to the total length of the ray. The projection
data gets updated after calculations through each ray. This procedure is referred to
as ART2. The individual steps of this algorithm are:

Calculate the total value of weight function ($W_{i\theta}$) along each ray as:

start: 1 For each projection angle (θ):
start: 2 For each ray ($i\theta$):
start: 3 For each cell (j):

$$W_{i\theta} = \sum_{j=1}^{N} w_{i\theta,j} \times w_{i\theta,j}$$

close: 3
close: 2
close: 1
start: 4 Start iterations (k):
start: 5 For each projection angle (θ):
start: 6 For each ray ($i\theta$):
Compute the numerical projection (Eq. 4.7). Calculate the correction as:

$$\Delta\phi_{i\theta} = \phi_{i\theta} - \tilde{\phi}_{i\theta}$$

start: 7 For each cell (j):
If $w_{i\theta,j}$ is nonzero then:

$$f_j^{\text{new}} = f_j^{\text{old}} + \mu \frac{\Delta\phi_{i\theta} \times w_{i\theta,j}}{W_{i\theta}},$$

where μ is a relaxation factor.
close: 7
close: 6
close: 5
Check for convergence as per Eq. 4.8.
close: 4 (k)

4.4.1.3 Gilbert ART

Gilbert [6] developed independently a form of ART known as SIRT (Simultaneous Iterative Reconstruction Algorithm). In SIRT, the elements of the field function are modified after all the corrections corresponding to individual pixels have been calculated. This route is referred to as ART3. The numerically generated projections are computed once for all the angles and gets updated only after the completion of calculations through all the rays. For each ray from all angles, all the cells are examined to look for those rays which pass through a particular cell. For each cell, the rays which pass through it will contribute a correction that is decided by the weight factor w_{ij}. The algebraic average of all these corrections is implemented on the cell. The individual steps of the algorithm are:

Calculate the total value of weight function ($W_{i\theta}$) along each ray as:

start: 1 For each projection angle (θ):

start: 2 For each ray ($i\theta$):

start: 3 For each cell (j):

$$W_{i\theta} = \sum_{j=1}^{N} w_{i\theta,j} \times w_{i\theta,j}$$

close: 3

close: 2

close: 1

start: 4 Start iterations (k):

start: 5 For each projection angle (θ):

start: 6 For each ray ($i\theta$):

Compute the numerical projection (Eq. 4.7)

Calculate the correction as:

$$\Delta\phi_{i\theta} = \phi_{i\theta} - \tilde{\phi}_{i\theta}$$

close: 6

close: 5

start: 7 For each cell (j):

Identify all the rays passing through a given cell (j). Let Mc_j be the total number of rays passing through the jth cell and corresponding $i\theta$, $w_{i\theta,j}$, $W_{i\theta}$ and $\Delta\phi_{i\theta}$. Apply correction as:

$$f_j^{\text{new}} = f_j^{\text{old}} + \frac{1}{Mc_j} \sum_{Mc_j} \mu \frac{w_{i\theta,j} \Delta\phi_{i\theta}}{W_{i\theta}},$$

where μ is the relaxation factor.

close: 7

Check for convergence as per Eq. 4.8.

close: 4(k)

4.4.1.4 Anderson ART

Anderson and Kak [1] proposed a variation to the ART algorithm. Abbreviated as SART (Simultaneous Algebraic Reconstruction Technique), the method of implementing the correction is similar to ART1. The difference from ART1 is in the calculation of correction for each cell. The weight factor used here is the exact intersection of a ray with the concerned cell. In contrast, ART1 uses the average correction for all the cells. This algorithm is referred to as ART4 in the following discussion. The individual steps of ART4 are:

Calculate the total value of weight function $(W_{i\theta})$ along each ray as:
start: 1 For each projection angle (θ):
start: 2 For each ray $(i\theta)$:
start: 3 For each cell (j):

$$W_{i\theta} = \sum_{j=1}^{N} w_{i\theta,j} \times w_{i\theta,j}$$

close: 3
close: 2
close: 1
start: 4 Start iterations (k):
start: 5 For each projection angle (θ):
start: 6 For each ray $(i\theta)$:
Compute the numerical projection (Eq. 4.7)
close: 6
start: 7 For each ray $(i\theta)$:
Calculate the correction as:

$$\Delta\phi_{i\theta} = \phi_{i\theta} - \tilde{\phi}_{i\theta}$$

start: 8 For each cell (j):
If $w_{i\theta,j}$ is non-zero then:

$$f_j^{\text{new}} = f_j^{\text{old}} + \mu \frac{\Delta\phi_{i\theta} \times w_{i\theta,j}}{W_{i\theta}}$$

where μ is a relaxation factor.
close: 8
close: 7
close: 5
Check for convergence as per Eq. 4.8.
close: 4(k)

4.4.2 MART

When the corrections in the iterative algorithms are multiplicative rather than additive, the algorithms are grouped under the family of MART (Verhoeven, [35]). Gordon [7] and Gordon and Herman [8] have suggested different forms of MART. The MART algorithms presented below are similar to those of [35].

The major difference between ART and MART algorithms is in the method of computing the corrections. While ART uses the difference between the calculated projections and measured projections, MART uses the ratio between the two. Hence the corrections applied to each cell during calculations are via the multiplication operation. The structure otherwise is similar to Gordon's ART (ART2).

The individual steps of three versions of MART (1, 2 and 3) are summarized below. *start: 1* Start iterations (k):
start: 2 For each projection angle (θ):
start: 3 For each ray ($i\theta$):
Compute the numerical projection (Eq. 4.7)
Calculate the correction as:

$$\Delta\phi_{i\theta} = \frac{\phi_{i\theta}}{\tilde{\phi}_{i\theta}}$$

start: 4 For each cell (j):
If $w_{i\theta,j}$ is nonzero then:
MART1:

$$f_j^{\text{new}} = f_j^{\text{old}} \times (1.0 - \mu \times (\Delta\phi_{i\theta}))$$

MART2:

$$f_j^{\text{new}} = f_j^{\text{old}} \times \left(1.0 - \mu \times \frac{w_{i\theta,j}}{(w_{i\theta,j})_{\text{max}}} \times (1.0 - \Delta\phi_{i\theta})\right)$$

MART3:

$$f_j^{\text{new}} = f_j^{\text{old}} \times (\Delta\phi_{i\theta})^{(\mu w_{i\theta,j}/(w_{i\theta,j})_{\text{max}})}$$

where μ is a relaxation factor.
close: 4
close: 3
close: 2
Check for convergence as per Eq. 4.8.
close: 1

Steps 3 and 4 form the essence of the reconstruction algorithm. All three versions include the relaxation factor μ. Typical values of the relaxation factor reported are in the range 0.01–0.1, larger values leading to divergence. It is to be noted that the correction calculated in step 3 is the ratio of the recorded projection data ($\phi_{i\theta}$) and that calculated from the guessed field, namely ($\tilde{\phi}_{i\theta}$) which is being iterated. The three versions of MART differ in the manner in which the corrections are implemented. In MART1, the weight function is prescribed in binary form, being unity if a particular ray passes through a cell and zero otherwise. In MART2 and MART3, the weight function is precisely calculated as the ratio of the length of the ray intercepted by the pixel and the maximum dimension of a segment enclosed by it.

4.4.2.1 AVMART

The reconstruction of a function from a finite number of projections recorded at different view angles leads to an ill-posed matrix inversion problem. The problem is accentuated when the projection data is limited. The resulting matrix is rectangular with the number of unknowns being greater than the number of equations. In view of the ill-conditioning of the matrix, the convergence of the iterations to any particular solution is dependent on the initial guess, the noise level in the projection data, and the under-relaxation parameter employed. The MART algorithm can be extended to (1) enlarge the range of the usable relaxation factor, (2) diminish the influence of noise in the projection data, and (3) guarantee a meaningful solution when the initial guess is simply a constant.

AVMART (AV for average) is a new approach to applying corrections by considering all the rays from all the angles passing through a given pixel [19]. Instead of a single correction obtained from individual rays, a correction that is determined as the average of all the rays is used. The difference between the conventional MART and the AVMART implementation is the following. The correction at each pixel is updated on the basis of the Nth root of the product of all the corrections from all the N rays of all view angles passing through a pixel. This idea is based on the fact that average corrections are expected to behave better in the presence of noisy projection data. Since an average correction is introduced, the algorithm is desensitized to noise. There is, however, a potential drawback. Since an average pixel correction based on a set of rays is computed, the reconstructed field is not required to satisfy exactly the recorded projection data. Validation studies did not show this weakness to be a cause for concern.

The important step of AVMART, namely step 4 is presented here, while all other steps remain unchanged.

start: 4 For each cell (j):

Identify all the rays passing through a given cell (j). Let Mc_j be the total number of rays passing through the jth cell.

Apply correction as:

AVMART1:

$$f_j^{\text{new}} = f_j^{\text{old}} \times \left(\prod_{Mc_j} (1.0 - \mu \times (\Delta\phi_{i\theta})) \right)^{1/Mc_j}$$

AVMART2:

$$f_j^{\text{new}} = f_j^{\text{old}} \times \left(\prod_{Mc_j} \left(1.0 - \mu \times \frac{w_{i\theta,j}}{(w_{i\theta,j})_{\text{max}}} \times (1.0 - \Delta\phi_{i\theta})\right) \right)^{1/Mc_j}$$

AVMART3:

$$f_j^{\text{new}} = f_j^{\text{old}} \times \left(\prod_{Mc_j} (\Delta\phi_{i\theta})^{\mu w_{i\theta,j}/(w_{i\theta,j})_{\max}} \right)^{1/Mc_j}$$

close: 4

The symbol $\prod$ in the three algorithms above represents a product over the index Mc_j. The Mc_jth root of this product is evaluated in each approach. The relaxation factor μ has been retained in the statements above for completeness, though it has been set as unity in calculations.

4.4.3 Maximum Entropy

Based on information theory, one can perform image analysis and construct meaningful tomographic algorithms. Suppose there is a source which generates a discrete set of independent messages r_k with probabilities p_k. The information associated with r_k is defined logarithmically as

$$I_k = -\ln p_k.$$

The entropy of the source is defined as the average information generated by the source and can be calculated as

$$\text{entropy} = -\sum_{k=1}^{L} p_k \ln p_k.$$

When the source is the image, the probability can be replaced by the gray level f_j, for the jth pixel and entropy can be redefined as

$$\text{entropy} = -\sum_{j=1}^{N} f_j \ln f_j.$$

For natural systems, the organization of intensities f_j over the image can be expected to follow the second law of thermodynamics namely,

$$f_j : -\sum_j f_j \ln f_j = \text{maximum}.$$

This is the basis of the MAXENT algorithm. For interferometric images, one can view the pixel temperature as the information content and entropy built up using their magnitudes. In the absence of any constraint, the solution of the above optimization problem will correspond to a constant temperature distribution, more generally a uniform histogram in terms of probabilities. Hence, the MAXENT algorithm is properly posed along with the projections as constraints.

Requiring that the entropy of the system be a maximum along with optical projections as constraints is known as the Maximum entropy optimization technique (MAXENT) [9]. It produces an unbiased solution and is maximally non-committal about the unmeasured parameters. This technique is particularly attractive when the projection data is incomplete [4] and is described below:

Consider a continuous function $f(x, y, z)$ with the condition $f(x, y, z) > 0$ and values f_j at $j = 1 \ldots N$ pixels. In the present context, the entropy technique refers to the extremization of the functional

$$F = - \sum_{j=1}^{N} f_j \ln |f_j| \tag{4.9}$$

subject to a set of constraints. In MAXENT the collected projection data and any other a priori information about the field to be reconstructed can be viewed as the constraints over which the entropy is to be maximized. A typical maximum entropy problem can be stated as:

$$\text{maximize} \left(- \sum_{j=1}^{N} f_j \ln |f_j| \right) \tag{4.10}$$

$$\text{subject to } \phi_i = \sum_{j=1}^{N} w_{ij} f_j$$
$$\text{and} \qquad f_j > 0.$$

Various schemes are available for extremizing a functional over some constraints, for example the Lagrange multiplier technique.

The maximum entropy algorithm has been shown to be equivalent to MART in the literature and is not considered further.

4.4.3.1 Minimum Energy

The MAXENT algorithm can be generalized for any other function in place of entropy. Gull and Newton [9] have suggested four such functions which can be maximized with the projections as constraints for tomographic reconstruction. After entropy, the energy functions are attractive and natural to use in physical problems. The minimum energy method (MEM) can be implemented in a manner analogous to MAXENT as follows:

$$\text{maximize} \quad \left(- \sum_{j=1}^{N} f_j^2 \right) \tag{4.11}$$

$$\text{subject to} \quad \phi_i = \sum_{j=1}^{N} w_{\text{ij}} f_j.$$

Compared to MAXENT, MEM has a simpler implementation while using the Lagrangian multiplier technique, since it results in a set of linear equations. Gull and Newton [9], however, have recommended MAXENT over MEM, since they found that the MEM produces a biased field that is negatively correlated.

4.5 Testing of Tomographic Algorithms

ART, MART, and the optimization algorithms were tested for variety of cases by Subbarao et al. [30] for synthetically generated temperature fields. Hence, it was possible to determine explicitly the convergence properties and errors for each of the methods. Among the various algorithms, the authors have identified MART3 as the best in terms of the error and CPU time requirements. The AVMART algorithms have been validated in the present section against two benchmark cases: 1. a circular region with 5 holes and 2. numerically generated 3D temperature field representing cellular fluid convection. Employing a temperature field similar to that encountered in the experiments aids in the choice of the proper initial guess and the error levels to be anticipated. This also helps in selecting the appropriate tomographic algorithm. Sensitivity of the algorithms to noise has been tested in the context of numerically generated temperature data. Issues addressed in the sensitivity study are initial guess, noise in projection data, and the effect of increasing number of projections on the accuracy of reconstruction. Iterative techniques for interferometric tomography have been discussed in [17–20].

4.5.1 Reconstruction of a Circular Disk with Holes

A circular region with five symmetrically placed holes is considered. The object is recognized in terms of the local dimensionless density, which is zero at the holes and unity elsewhere. To implement the reconstruction algorithm, it is convenient to enclose the circular object within a square domain. The gap between the circle and the square is specified to have zero density (in calculations, a value of 0.001 has been used for zero density). The square region is discretized into 61×61 cells in the x and y directions. Projections of the object are determined analytically and are exact. The recovery of the original object from a limited number of these projections using the original MART as well as the proposed AVMART algorithms is discussed below.

Projections at angles of 0, 45, 90, and 135° are considered in the present application. The initial guess for the density field was a constant value of unity. A convergence criterion of 1 % for the iterations is uniformly used. At a convergence of 0.01 %,

Fig. 4.3 Original and reconstructed density fields of a
circular region with holes (the
outer circle appears as an octa-
gon because of a finite number
of view angles employed). The
right column presents reconstructions with AVMART
(1–3)

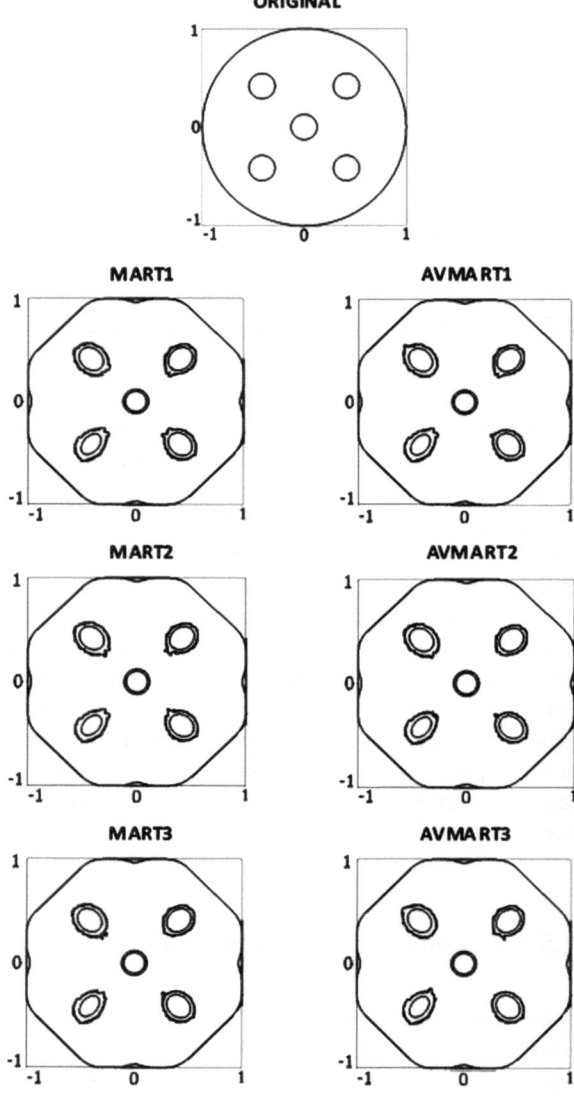

the solution was practically identical, except that errors were seen to be marginally
higher. This feature of tomographic algorithms, that convergence is asymptotic (but
not monotonic) has been reported for rectangular systems of linear algebraic equations. Such trends are to be expected in the reconstruction of fields having a step
discontinuity, at the hole boundary in the present example. The relaxation factor was
set at 0.1 in case of original MART while it was unity in the AVMART algorithms.

Table 4.1 Comparison of the MART and AVMART algorithms for a circular region with holes

Quantity	MART1	MART2	MART3	AVMART1	AVMART2	AVMART3
E_1	0.99	0.96	0.95	0.99	0.96	0.96
E_2	0.25	0.24	0.23	0.24	0.23	0.23
$E_3,\%$	25.12	24.08	23.63	24.59	23.72	23.65
Number of points (%) having error in the range						
>95%	0.27	0.05	0.05	0.27	0.05	0.05
75–95%	0.64	0.62	0.86	0.83	0.72	0.70
50–75%	3.90	4.11	4.43	3.47	4.00	3.98
Iterations	51	63	29	17	24	21
CPU (minutes)	9.51	11.97	5.65	0.32	0.45	0.40

A summary of the reconstructed fields using the three original and three proposed algorithms is shown in Fig. 4.3. In principle, all the six algorithms converge to a qualitatively meaningful solution. The void fraction, namely the fraction of the space occupied by the holes was 0.34 in the present example. In the reconstructed solution, the void fraction can be determined from the formula

$$\text{void fraction} = 1 - \frac{\sum_{i=1}^{N} \rho_i}{N}.$$

It was found that all the six algorithms reproduced precisely a void fraction of 0.34. The algorithms, however, differed in terms of CPU time, errors, and error distribution. The three different error norms reported in the present work are:

$$E_1 = \max[\text{abs}(\rho_{\text{orig}} - \rho_{\text{recon}})], \quad \text{maximum of absolute difference,}$$

$$E_2 = \sqrt{\frac{\sum[(\rho_{\text{orig}} - \rho_{\text{recon}})]^2}{N}}, \quad \text{RMS error,}$$

$$E_3 = \frac{E_2}{\rho_{\text{max}} - \rho_{\text{min}}} \times 100, \quad \text{normalized RMS error, \%}$$

Results for the error level distribution in the reconstructed field have also been determined. The distribution of the absolute error as a percentage of the E_1 error has been presented in the regions >95%, 75–95%, and 50–75%.

Errors and their distribution along with the computational details are given in Table 4.1. These errors are not visible in Fig. 4.3 where the data has been smoothed before drawing. It is clear from Table 4.1 that the errors of all the six algorithms are practically close, with those of MART1 and AVMART1 being marginally on the higher side. An examination of the error distribution shows that large errors (>95%) are seen only over 0.27% of the physical region. Specifically, large errors

Fig. 4.4 Temperature surface over the midplane of the fluid layer, in the form of cubic cells. This data is used for calculation of the projection data and subsequently for the assessment of reconstruction errors

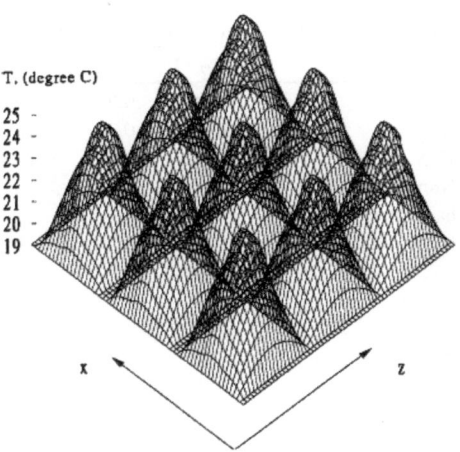

are restricted to the surface of the holes, where a step change in the field property (the density in the present example) takes place. The errors are uniformly small elsewhere. The most significant difference between the original and the proposed algorithms is in terms of the number of iterations (and correspondingly, the CPU time). The proposed algorithms require fewer iterations for convergence and require a smaller CPU time. This is clear evidence of the computational efficiency of these algorithms in the context of exact projection data.

4.5.2 Reconstruction of a Numerically Generated Thermal Field

The second application taken up for analysis comprises a numerically generated convective thermal field in a horizontal differentially heated fluid layer. For definiteness, the wall temperatures employed are 15 and 30°C, respectively. The 3D temperature field has been determined as follows. The stream function, vorticity, and energy equations are solved in two dimensions with symmetry conditions applied on the side walls, by a finite difference method. The solution thus obtained corresponds to a system of longitudinal rolls spread over an infinite fluid layer. Such geometries show a polygonal planform corresponding to a fully 3D temperature field [21]. Three dimensionality has been simulated in the present work by superimposing a sine variation in the thermal field parallel to the axis of the roll. A surface plot of the resulting temperature field revealed the flow to be organized in the form of cubic cells in the fluid layer (Fig. 4.4).

The advantages of selecting the field to be reconstructed in the particular manner outlined above are: 1. The field is continuous and hence reconstruction errors can be expected to be small, as compared to the application with holes. 2. Errors with perfect data being small, one can systematically study errors induced by the initial guess

Table 4.2 Performance comparison of the AVMART algorithms in a differentially heated fluid layer

Initial guess	Quantity	AVMART1	AVMART2	AVMART3
Constant	E_1, °C	1.97	1.97	1.97
	E_2, °C	0.49	0.48	0.49
	E_3, %	2.86	2.79	2.86
	Iterations	9	12	14
	CPU (sec)	30.6	41.3	47.2
Two dimensional longitudinal rolls	E_1, °C	1.98	1.98	1.98
	E_2, °C	0.49	0.49	0.49
	E_3, %	2.86	2.86	2.86
	Iterations	8	12	12
	CPU (sec)	28.9	41.2	42.7
Random	E_1, °C	12.15	13.42	6.20
	E_2, °C	5.59	4.74	0.60
	E_3,%	32.70	27.77	3.50
	Iterations	15	17	14
	CPU (sec)	52.8	59.1	47.8

and noise in the projection data. 3. The thermal field being analyzed is physically realizable.

For reconstruction, the fluid layer has been discretized into 11 planes and each plane into 61 × 61 cells. The relaxation factor in AVMART is set to unity. Since the algorithms are being tested under conditions of limited data, only two and four projections have been considered. A convergence criterion of 0.01 % has been uniformly employed in the computation. The errors reported here are on the basis of the entire fluid layer. In terms of temperature, the three different errors reported are:

$$E_1 = \max[\text{abs}(T_{\text{orig}} - T_{\text{recon}})], \quad \text{maximum of absolute difference, ° C}$$

$$E_2 = \sqrt{\frac{\sum[(T_{\text{orig}} - T_{\text{recon}})]^2}{N}}, \quad \text{RMS error, ° C}$$

$$E_3 = \frac{E_2}{(T_{\text{hot}} - T_{\text{cold}})} \times 100 \quad \text{normalized RMS error, %.}$$

In these definitions, T_{hot} and T_{cold} are the hot and cold plate temperatures. T_{orig} and T_{recon} are the temperature variables of the original and the reconstructed field respectively. Results for the error level distribution in the fluid layer have also been determined. The distribution of the absolute error as a percentage of the E_1 error has been presented in the three zones, namely, >95 %, 75–95 %, and 50–75 %.

Table 4.3 Fractional distribution of the E_1 error over the fluid layer

Initial guess	Number of points (%) having error in the	AVMART1	AVMART2	AVMART3
(1)	>95 %	0.17	0.17	0.17
	75–95 %	0.57	0.48	0.57
	50–75 %	5.76	5.15	5.73
(2)	>95 %	0.17	0.17	0.17
	75–95 %	0.60	0.62	0.62
	50–75 %	5.68	5.58	5.58
(3)	>95 %	0.02	0.01	0.002
	75–95 %	5.79	2.00	0.02
	50–75 %	34.46	11.92	0.30

4.5.2.1 Sensitivity to Initial Guess

The inversion of matrices arising from the ART family of algorithms from limited projection data is a mathematically ill-posed problem. As a rule, the number of equations here is smaller than the number of unknowns. This makes the solution-set infinite in the sense that a unique solution is not guaranteed. Different initial guesses may in principle, lead to distinct solutions of this infinite set. In the absence of any knowledge about the field being studied, it is a difficult task to prescribe the initial guess. The sensitivity of the algorithms to the initial guess has been studied with reference to three different fields, namely:

1. A constant temperature field ($=1°C$)
2. Temperature distribution corresponding to 2D longitudinal rolls and
3. A random field between 0 and $1°$ C with an RMS value of $0.5 °C$.

The initial guesses 1 and 2 were seen to qualitatively reproduce the thermal field of Fig. 4.4 quite well. The reconstructed thermal field is not shown as it is very close to the original. The noise present in the third guess was seen to be present in the reconstructed data. But the noise could be filtered in the frequency domain using a band-pass filter. The reconstructed field after noise removal was seen to be similar to the original in Fig. 4.4. The errors, number of iterations, and the CPU time for the three initial guesses are presented in Table 4.2. The fractional distribution of errors are reported in Table 4.3. With initial guesses 1 and 2, the RMS, and fractional errors can be seen to be small for all the three algorithms. The maximum error is larger, but with reference to Table 4.3, it can be seen that large errors are restricted to small areas and are hence localized. Thus, in effect the initial guesses 1 and 2 may be considered to be equivalent. The errors corresponding to the third guess are uniformly higher for AVMART1 and AVMART2 algorithms, but small for AVMART3. The number of iterations for AVMART3 are also smaller. Hence, AVMART3 emerges as the best algorithm among those proposed in terms of errors and CPU time for a noisy initial guess.

Table 4.4 Comparison of AVMART algorithms when 5 % Noise is present in the projection data; 2-view reconstruction

Quantity	AVMART1	AVMART2	AVMART3
E_1, °C	4.452	4.449	4.450
E_2, °C	1.08	1.08	1.08
E_3, %	6.37	6.36	6.37
Number of points (%) having error in the range			
>95 %	0.004	0.004	0.004
75–95 %	0.222	0.200	0.222
50–75 %	4.400	4.387	4.400
Iterations	9	12	14
CPU (sec)	30.5	40.9	47.8

The insensitivity of AVMART3 algorithm to noise can be explained as follows. In the other two algorithms, correction is applied by finding the Mc_jth root of the product of all corrections arising from Mc_j rays. In the third, the root is corrected for the length of the intercept of each ray with the cell under question. This improves the estimate of the path integral.

4.5.2.2 Sensitivity to Noise in Projection Data

In measurements involving commercial grade optical components and recording and digitizing elements, the projection data is invariably superimposed with noise. Software operations such as interpolation and image processing can also contribute to errors in the projection data.

The performance of AVMART(1-2) algorithms are compared with noisy projection data as the input. A 5 % noise level has been adopted for all calculations. The noise pattern has been generated using a random number generator, with a uniform probability density function. Results have been presented for 2 and 4 projections corresponding to view angles of (0, 90°) and (0,60, 90, 150°), respectively. The initial guess for reconstruction with 2 projections is simply a constant; for 4 projections, the result obtained with 2 projections has been used as the initial guess.

Results with 2 projections show that all the three algorithms reproduce qualitatively the temperature field of Fig. 4.4. However, quantitative differences are to be seen. The noise level in the reconstructed field is found to be slightly higher than that in the projection data. The magnitude of the three different errors and the distribution of the fractional error over the fluid domain are presented in Table 4.4. All the three algorithms are practically equivalent in terms of errors, though AVMART2 is seen to be marginally better from the error point of view. However, the CPU time of AVMART1 is the smallest. It is to be noted that noise (in terms of E_3) in the projection data has been amplified during the reconstruction process (from 5 to 6.4 %). This

Table 4.5 Comparison of AVMART algorithms: 5 % noise inprojection data, 4-view reconstruction

Quantity	AVMART1	AVMART2	AVMART3
E_1, °C	11.80	5.52	5.52
E_2, °C	1.78	1.36	1.36
E_3, %	10.41	8.00	8.00
Number of points (%) having error in the range			
>95 %	0.004	0.007	0.007
75–95 %	0.029	0.349	0.346
50–75 %	0.276	5.186	5.177
Iterations	190	53	53
CPU (sec)	1,767.7	502.3	520.8

is in contrast to noise in the initial guess, where iterations tend to diminish errors in the converged field.

Reconstruction with 4 view angles is taken up next. Table 4.5 shows the error levels in the reconstructed data and the distribution of these errors within the fluid layer. It can be seen immediately that the E_3 errors with 4 projections are larger than those for 2 projections alone. The distribution of errors show that these are at best localized, i.e l arge errors may occur at a few sporadic points. The AVMART1 algorithm shows a considerable deterioration in performance, since errors as well as CPU time are substantially higher. AVMART2 and AVMART3 algorithms are seen to perform better than AVMART1. AVMART2 is marginally superior to AVMART3 since the error magnitudes are equal, but the former takes a smaller CPU time. Hence, a consolidated view to emerge from the discussion above is that AVMART2 exhibits the best performance.

It is of interest to compare the best proposed algorithm, namely AVMART2 with the best original MART algorithm identified by Subbarao et al. [30], namely MART3 of the present study. To this end, reconstruction was carried out using 2-views of 0 and 90° for convection in a horizontal differentially heated fluid layer, leading to 2D longitudinal rolls. The projection data was superimposed with 5 % noise and an initial guess of a constant temperature field was used. Errors for MART3 were seen to be amplified by a factor of 4 compared to a factor of 1.6 for AVMART2. The computer time was also higher by a factor of 4 when compared with AVMART2. However, the fractional distribution of errors over the fluid layer were seen to be similar for both, thus confirming that they continued to belong to the same family of algorithms.

The following inferences can now be drawn from the discussion above:

1. The three AVMART algorithms show similar performance in the presence of noise in the projection data. AVMART2 is, however, marginally superior in terms of errors and CPU time.
2. The noise in the projection data persists after reconstruction.
3. Increasing the number of noisy projections amplifies the error in reconstruction.

Fig. 4.5 Definition of partial projection data in the context of crystal growth; also see Fig. 5.9

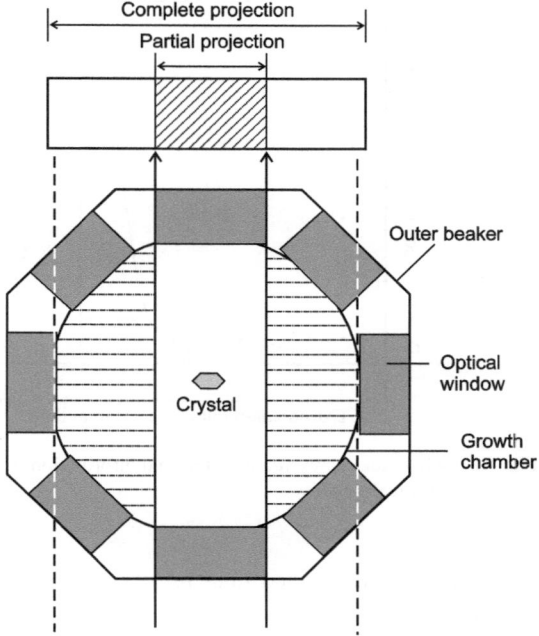

4. AVMART2 clearly shows superiority over MART3 for noisy projection data. Hence, it supercedes MART3 as the favored tomographic algorithm for the class of problems studied.

4.6 Extrapolation Scheme

Tomographic reconstruction requires each projection to span the entire object. Meeting this requirement can be quite difficult in real systems where the size of the light beam could be smaller than the test object. An extreme example is found in crystal growth, Sect. 5.2, where the beaker containing the aqueous solution is 160 mm in diameter while the light beam itself is 40 mm diameter, being limited by the diameter of the optical windows. The projection data is thus incomplete, as shown schematically in Fig. 4.5.

In order to successfully apply tomographic algorithms for 3D reconstruction, one needs projection data over the entire width of the physical domain in every view angle. In this respect, the experimentally recorded partial projection data can be extrapolated to derive information about the portion of the physical domain beyond the light beam. The extrapolation scheme is schematically shown in Fig. 4.6 where the solid curve represents the partial experimental data as retrieved from the recorded

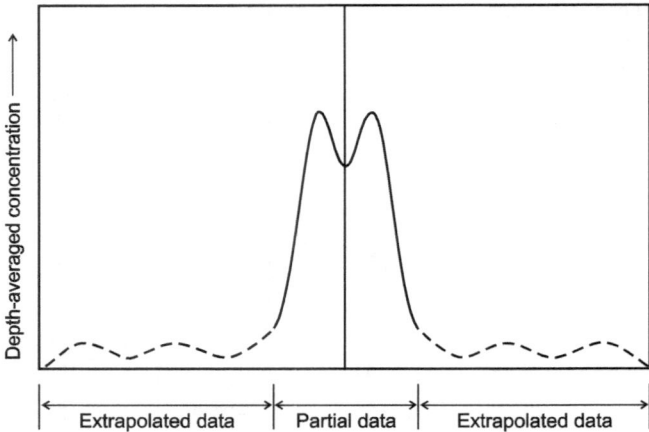

Fig. 4.6 Schematic representation of partial information from optical images and extrapolated data beyond the optical window

optical images and dotted lines represent the extrapolated information beyond the optical windows.

Extrapolation methods are expected to succeed if the laser beam is arranged to record all important aspects of the process, as in convection around the crystal (Sect. 5.2). The data outside the measurement volume is then of secondary importance and can be estimated by extrapolation. Often, the field being studied is quite continuous for curve-fitting techniques to apply. Additional advantages would be accrued if the apparatus is a beaker since the path length of the light beam (namely, the chord length of the beaker) progressively decreases toward its edges. Hence, the overall reconstruction process is de-sensitized to the extrapolation step.

Referring to the crystal growth experiment, the property under question salt concentration in water. In the present work, a tenth order polynomial has been used to extrapolate the concentration distribution, starting with the portion covered by the optical windows. Polynomials of order 5–10 produced practically identical results. The limiting values of concentration in the far field, and the necessity of maintaining slope-continuity in the concentration distribution at every point have been enforced. An independent check on the accuracy of the experiment, data analysis, and extrapolation is the conservation of solutal mass in each of the projections. Mass balance was found to be better than 0.01 % in all the experiments analyzed using tomography. The above approach is first tested in the context of numerically simulated data for buoyancy-driven convection and is discussed in the following section.

4.7 Validation of Reconstruction Procedure with Simulated Data

Tomographic algorithms are first validated against numerically simulated data. Complete as well as partial data sets are considered, while the partial data set is dealt with using extrapolation. The physical problem considered for generating the simulated data is buoyancy-driven convection in a differentially heated circular fluid layer. The upper and lower walls are maintained at specified temperatures, and the sidewall is thermally insulated. The fluid considered is air and the Rayleigh number based on the height of the fluid layer is set at Ra = 12, 000. The temperature distribution in the fluid layer has been obtained by numerically solving the governing equations of flow and energy transport on a fine grid. For definiteness, the thermal field is taken to be axisymmetric; accordingly the isotherms on individual planes of the fluid layer are circular.

With temperature determined numerically, the projection of the thermal field is obtained by path integration. The projection data is inverted using tomographic algorithms and errors are explicitly calculated. The radial data set obtained by numerical simulation is transformed onto a rectangular grid. The grid dimensions depend upon the extent of partial data to be considered for error analysis. Full data, 60 % and 30 % data have been studied.

Errors have been reported in the present section on three grids, namely 64 × 64, 128 × 128, and 256 × 256. Here the first number represents the number of view angles along which projections have been recorded, and the second term indicates the number of rays for each view. The definitions of errors considered are:

$$E_1 = \max \left[(T_{\text{orig}} - T_{\text{recon}}) \right], \qquad (4.12)$$

$$E_2 = \sqrt{\frac{1}{N} \sum \left[T_{\text{orig}} - T_{\text{recon}} \right]^2}. \qquad (4.13)$$

Here T_{orig} and T_{recon} are the original and reconstructed temperature fields respectively and N is the total number of grid points on the reconstructed planes. All temperatures generated by simulation are dimensionless and in the range of 0 to unity. Equation 4.12 denotes the absolute maximum temperature difference and Eq. 4.13 represents the RMS error in the reconstructed temperature field. The difference between error norms E_1 and E_2 arises from the fact that the former highlights large isolated errors, while the latter reveals trends that are applicable for the entire cross-section.

The physical realizability of axisymmetric convection in a circular cavity around a Rayleigh number of 12,000 in the form of concentric rings has been discussed at length by Velarde and Normand [34] for a step change in the cavity temperature difference. Figure 4.7a shows a schematic diagram of concentric roll patterns in an axisymmetric fluid layer. Figure 4.7b shows numerically generated isotherms in the direction along the diameter of the test cell. An experimental evidence of this pattern is shown in Fig. 4.7c where concentric rolls are seen at Ra = 5, 861 when the viewing axis is at right angles to that of the circulation pattern within the rolls. Fig. 4.8 shows numerically generated projection data in the form of isotherms along the

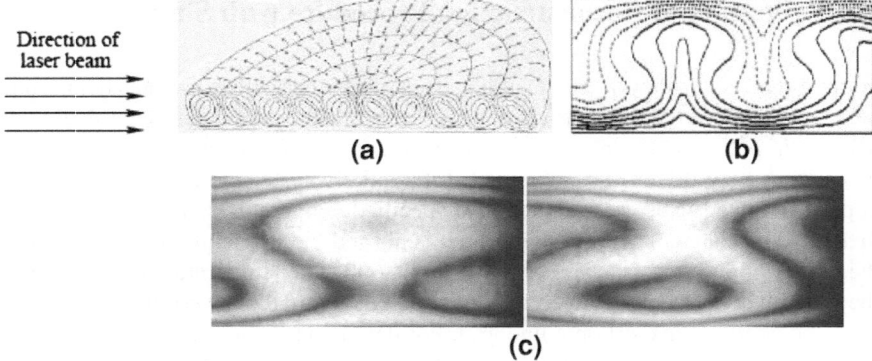

Direction of
laser beam

(a)   **(b)**

(c)

Fig. 4.7 a Visualization of concentric roll patterns in an axisymmetric fluid layer; **b** numerically
generated isotherms on a plane of a circular fluid layer appropriate for the roll patterns seen in **a**;
c an interferogram of convection in a circular fluid layer showing the formation of omega-shaped
rolls in a full projection [viewing direction is along the laser beam of **a**]

Fig. 4.8 Complete projection
data in the form of isotherms
for the differentially heated
circular fluid layer

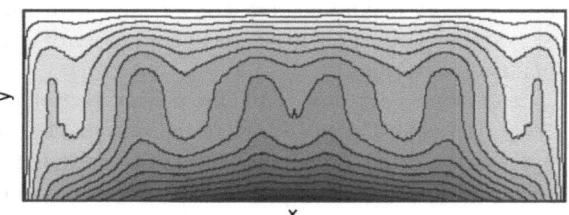

viewing direction. The data is presented in the form of contours of the path integrated
temperature field. Since the thermal field is axisymmetric, the projection data for all
other view angles are identical to Fig. 4.8.

Figure 4.9 shows the reconstruction over a horizontal plane of the fluid layer at
$y/H = 0.65$, where H is the height of the fluid layer and y represents the vertical
coordinate. It shows the reconstructed temperature field with complete projection
data (100 %) and partial data (60 % in Fig. 4.9b and 30 % in Fig. 4.9c), symmetrically placed about the center. The axisymmetric nature of temperature distribution is
brought out in all the reconstructions. They are in agreement with the experimental
results presented in Fig. 4.9d that shows the presence of concentric rings at a given
plane of the circular fluid layer when visualized from the top. This can be taken as a
validation of the extrapolation procedure used to convert partial to an approximate
but a complete data set.

A quantitative comparison of the reconstructed temperature profiles along the
diameter of the cavity for different combinations of rays and views is shown in
Fig. 4.10. Profiles obtained with full as well as partial data are reported. For the
complete data set, a perfect match between the original and reconstructed profiles
can be seen for grid sizes of 128×128 and 256×256, while small errors are to

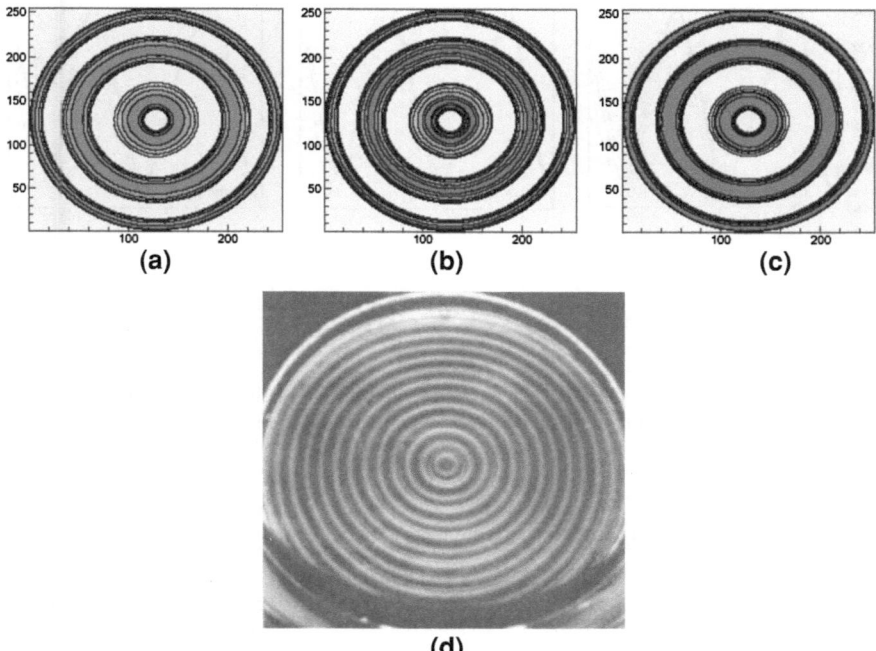

(a)   (b)   (c)

(d)

Fig. 4.9 Reconstructed temperature contours at $y/H = 0.65$ for full **a** 100 % and partial [60 % **b**, 30 % **c**] projection data; **d** the possibility of the presence of concentric rings in an axisymmetric fluid layer is shown when viewed from the top

be seen for the 64×64 grid. The extent of deviation from the original increases as the fraction of incomplete data increases. Noticeable errors are to be seen when only 30 % of the original data is used, the rest of it being derived by extrapolation. Errors in reconstruction were found to be significantly higher when the partial data set was used without extrapolation.

The magnitude of errors as a function of discretization of the fluid layer and size of the partial data set are summarized in Table 4.6. Since the difference between the minimum and maximum temperatures is unity, the percentage error is obtained as $100 \times E_1$ and $100 \times E_2$. In Table 4.6, error E_1 is consistently seen to be higher that E_2, the latter being an average over the entire field. Both errors decrease as the grid size (number of rays and views) increases. For a given grid size, errors increase as the fraction of original data used in the reconstruction decrease. When only 30 % of the original data is used (the rest being obtained by extrapolation), the maximum errors on a 64×64 grid are 17.3 % (absolute maximum) and 7.6 % (RMS). Figure 4.9 shows the corresponding reconstructions are qualitatively meaningful, and hence these error magnitudes may be taken to be within limits.

In the above discussion pertaining to validation, an axisymmetric temperature field has been considered. Hence projections generated from all the view angles

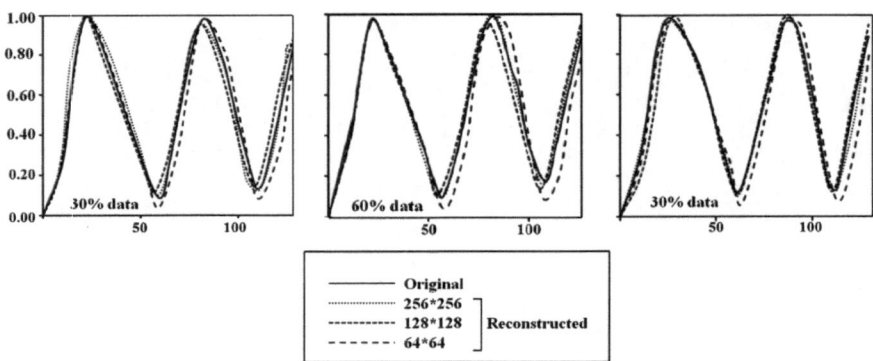

Fig. 4.10 Comparison of original and reconstructed temperature distribution along the radial direction for three different combinations of rays and views; **a** 100% and partial [60% **b**, 30% **c**] projection data

Table 4.6 Comparison of the original and reconstructed temperature fields in terms of errors E_1 and E_2 for buoyancy-driven convection in a circular cavity

Data type	Rays × Views	E_1	E_2
	256 × 256	0.052	0.028
Full data	128 × 128	0.109	0.056
	64 × 64	0.124	0.058
	256 × 256	0.095	0.039
60% Data	128 × 128	0.148	0.061
	64 × 64	0.152	0.067
	256 × 256	0.122	0.051
30% Data	128 × 128	0.148	0.072
	64 × 64	0.173	0.076

are identical to that shown in Fig. 4.8. In order to check the feasibility of the overall reconstruction procedure in more general terms, an unsymmetric distribution of temperature was generated by multiplying the numerically simulated temperature field with a function which depends upon the view angle. This function was chosen as $(1 + \alpha \sin \theta)$ where θ represents the view angle and α is a constant. Hence, the projection data now depends upon the view angle. Figure 4.11 shows the reconstructed temperature fields at a plane located at $y/H = 0.65$ for two values of α (=0.1 and 0.2). A clear breakdown in the axisymmetric nature of the flow field can be seen from the reconstructed temperature distribution as compared to that shown in Fig. 4.9.

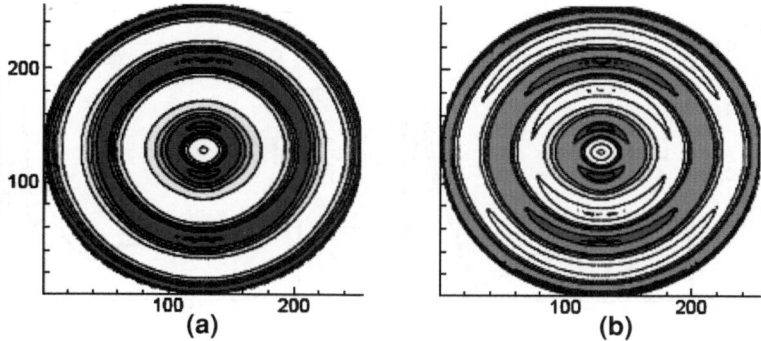

Fig. 4.11 Reconstructed temperature contours for a fluid layer which is not strictly axisymmetric for the two values of α (=0.1 **a** and 0.2**b**)

4.7.1 Comparison of ART and CBP for Experimental Data

Imaging of convection patterns around the growing KDP crystal is discussed here. Four view angles are considered for tomographic reconstruction of the concentration field. In the experiments discussed, the crystal is not disturbed during the recording of the projection data. In addition, longer durations of crystal growth have been considered. Since finite time is required to turn the beaker and record projections, experiments have been conducted when the concentration field around the crystal is nominally steady. Occasionally, the plume above the crystal is marginally unsteady, and a time-averaged sequence of schlieren images has been used for analysis. The increment in the view angles is 45° covering the range of 0–180°.

The crystal growth process is initiated by inserting a spontaneously crystallized KDP seed into its supersaturated aqueous solution at an average temperature of 35 °C. This step is followed by slow cooling of the aqueous solution. The cooling rate employed is 0.05 °C/h. The seed thermally equilibrates with the solution in about 20 min. With the passage of time, the density differences within the solution are solely due to concentration differences. Adjacent to the crystal, the solute deposits on the crystal faces, and the solution goes from the supersaturated to the saturated state. When the crystal size is small, salt deposition from the solution onto the crystal occurs by gradient diffusion. Diffusive transport has been sustained for a longer duration of time (and larger crystal sizes as well) by maintaining a small degree of supersaturation in the solution. For larger crystals, the denser solution in the beaker displaces the lighter solution close to the crystal, and a circulation pattern is setup. The onset of fluid motion is determined by the relative magnitudes of the driving buoyancy force and the resisting viscous force. The buoyant plume resulting from fluid motion is essential for transporting the solute from the bulk of the solution to the crystal and determines the crystal growth rate at later stages of growth process. The plume is visible as the spread of light intensity in the schlieren images.

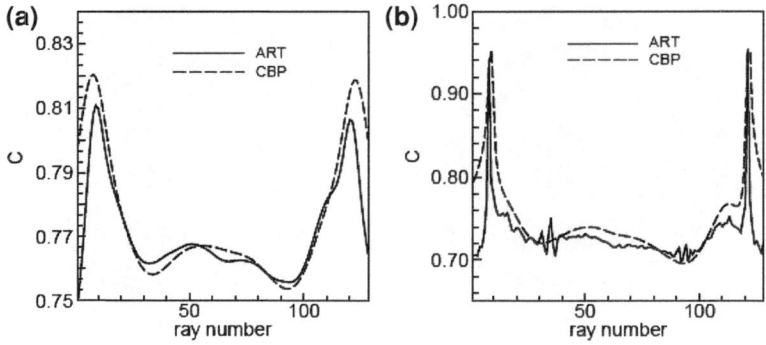

Fig. 4.12 Comparison of concentration profiles along a sector passing through the vicinity of the growing crystal using ART and CBP. Two horizontal planes above the growing crystal ($y/H = 0.05$ and 0.30) are considered

Reconstructions based on ART and CBP are now discussed. Four projections of schlieren images are recorded by turning the growth chamber while keeping the crystal stationary. Complete information for the full width of the growth chamber is generated using the extrapolation scheme discussed earlier, whereas linear interpolation is employed to generate intermediate view angles (in the range of 0–180°). Figure 4.12 shows the reconstructed profiles using ART and CBP along a sector passing through the vicinity of the growing crystal for two horizontal planes, $y/H = 0.05$ and 0.30 above the crystal. A good overall match between the two reconstruction approaches can be seen. As a rule, the computational time for ART is, however, much higher than CBP.

4.8 Jet Interactions

Tomographic reconstruction of density variation in mixing jets using color schlieren images is presented. The two jets studied are both of helium flowing into the normal ambient. The nozzle orientation of the two jets is vertically downwards. Buoyancy effects are created by helium, the lighter gas. The dimensionless parameters of the flow configuration, namely, Reynolds and Froude numbers are given in Table 4.7. For the range of parameters studied, the flow pattern is steady, permitting the record of optical projections. The color schlieren images are provided at an angular interval of 90° though images were recorded at 10° intervals. Reconstruction has been carried out using the CBP algorithm.

Figure 4.13 shows schlieren images and their subsequent 3D reconstruction in terms of the vertical density gradient for the parameters listed in Table 4.7. The volumetric reconstruction of jets shows their overall structure, spreading, and depth of penetration in the vertical direction. Stronger buoyancy effect (measured by Froude

Table 4.7 Flow parameters considered in color schlieren imaging of helium–oxygen jets

Figure number		Gas	Reynolds number	Froude number
	Jet A	Helium	47.43	4.88
	Jet B	Helium	47.43	4.88
	Jet A	Helium	111.23	11.39
	Jet B	Helium	31.61	3.25
Figure 4.13	Jet A	Helium	111.23	11.39
	Jet B	Helium	47.43	4.88

number) increases spreading through transverse diffusion and reduces the depth of penetration. Larger jet momentum, and hence Reynolds number, increases the depth of penetration but the jet is prone to instability. When pairs of jets are used, their individual regimes can lead to complex patterns of density distribution.

4.9 Treatment of Unsteady Data

In a convection experiment, the projection data for various view angles can be recorded either by turning the experimental chamber or the source-detector combination. A finite time elapses during measurements and the projection data recorded is asynchronous in time. When the field of interest varies with time, the projection data from all view angles should be correlated for principles of tomography to apply. This method of recording the projection data requires multiple pairs of source-detector combinations. The technique of proper orthogonal decomposition (POD) applied to the projection sequence offers a simpler alternative to handle asynchronous projection data. The central idea depends on the modes being ordered though the data collected is asynchronous.

Proper orthogonal decomposition, also known as Karhunen–Loeve decomposition, starts with an ensemble of data, called snapshots, collected from an experiment [25]. The technique extracts mode shapes or basis functions optimally, providing an efficient way of capturing the dominant components of a continuous process with a finite number of modes. Though well-established in image processing and pattern recognition, its application in measurements of transport phenomena is limited. Sirovich [25] introduced the method of snapshots to efficiently determine the POD eigenfunctions for large problems. It has been widely applied to unsteady computational fluid dynamics formulations. Torniainen et al. [33] used POD as the basis for the analysis of unsteady reacting flows imaged by holographic interferometry. The present discussion is concerned with the application of the POD technique jointly with either CBP or any of the iterative tomographic algorithms. The study is conducted for the spatial reconstruction of time-varying density gradient fields when the flow field is imaged by a schlieren system.

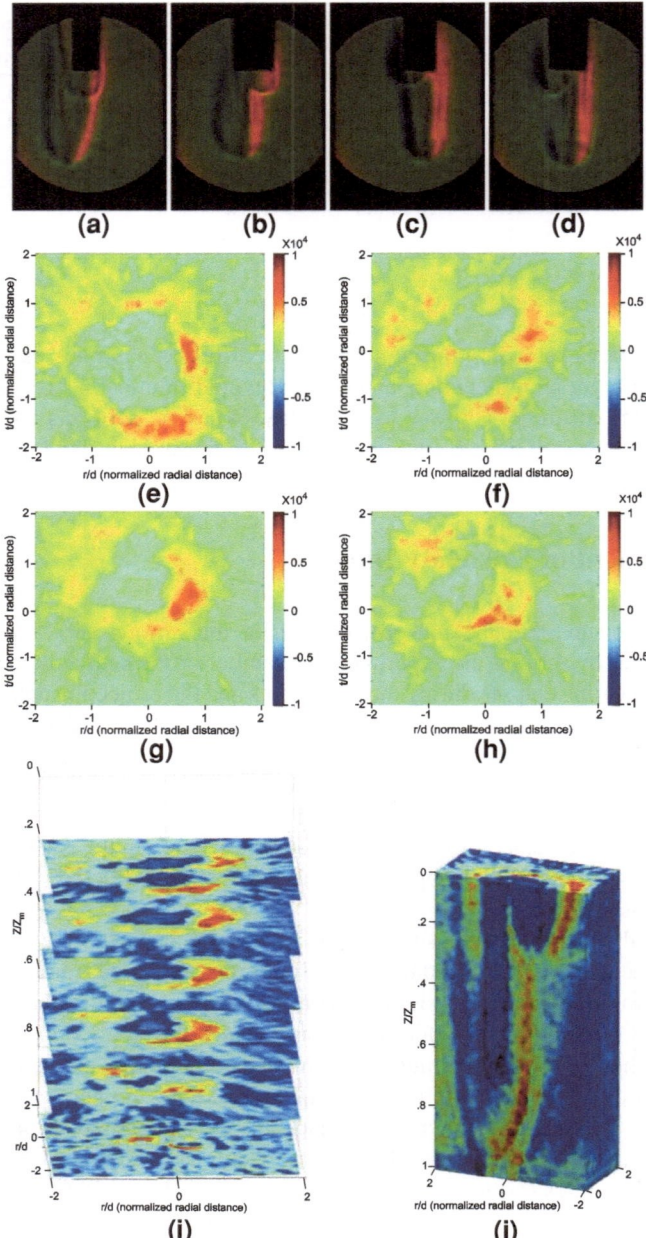

Fig. 4.13 Colour schlieren images visualized along **a** 0°; **b** 90°; **c** 180°; **d** 270°; reconstructed density gradient information at **e** $z/z_{ml} = 0.5$, **f** 0.9, **g** 0.5, **h** 0.9, **i** 3D slice of the reconstructed cross-sections, and **j** Volumetric representation of flow

The central idea behind the proposal is the following: By definition, optical projections of a given flow field in the schlieren arrangement are line integrals of the density gradient field along the viewing direction. In POD, the gradient field is decomposed into a product of two functions that individually depend on space and time. The path integrals are only defined in space. Hence, the time-dependent function must remain unaltered even though projections are recorded at different time instants. The aim of completely decoupling the temporal and spatial components of the field of interest is thus accomplished using proper orthogonal decomposition. The integration of POD with tomography presents a novel approach to reconstruct the unsteady field of interest from its asynchronous projection data.

The basic POD procedure can be summarized as follows: An ensemble of images collected over a period of time can be represented by the symbol $\psi(x, t)$ and approximated as a finite sum in the variables separated form

$$\psi(x, t) \approx \sum_{k=1}^{N} u_k(t) v_k(x). \tag{4.14}$$

The approximation becomes exact as N approaches infinity. In Eq. 4.14, symbol t is the time index when the image is recorded and x is the pixel location vector in 1 or 2 dimensions. The image definition for the present study is in terms of a set of intensity values at various pixel locations recorded at a time instant t. The representation of ψ in terms of the basis functions is not unique. Several choices of $v_k(x)$ are possible and for each choice the sequence of time-dependent functions $u_k(t)$ is unique. Proper orthogonal decomposition is concerned by finding the best possible choice of the functions $v_k(x)$ for a collection of images $\psi(x, t)$. The basis functions are taken to be orthonormal so that the determination of the coefficient function $u_k(t)$ for a given k will depend only on $v_k(x)$ and not on the other components. Orthonormality requires

$$\int_x v_k(x) v_l(x) \mathrm{d}x = 1 \quad \text{if} \quad k = l \quad \text{and } 0 \quad \text{otherwise.} \tag{4.15}$$

The time-dependent part can then be obtained as:

$$u_k(t) = \int_x \psi(x, t) v_k(x) \mathrm{d}x. \tag{4.16}$$

The determination of the coefficient function $u_k(t)$ for a given k depends only on $v_k(x)$ and not on others. The orthonormal property of Eq. 4.15 is useful while selecting functions $v_k(x)$ only in the limit of N approaching infinity. If N is finite but reasonably large, we choose the basis functions $v_k(x)$ in such a way that the approximation for each N is the best possible in a least squares sense. These calculations can be carried out using commercial software such as MATLAB. Ordered, orthonormal functions

$v_k(x)$ are the proper orthogonal modes of $\psi(x, t)$. With these functions computed, Eq. 4.14 is the proper orthogonal decomposition of the image data $\psi(x, t)$.

The approach integrating tomography and proper orthogonal decomposition for 3D reconstruction of the unsteady concentration gradient field is described below. The steps involved are:

1. Record the time-dependent schlieren images for a given position of the laser-camera axis with respect to the test apparatus.
2. Turn the apparatus through an angular increment and record the time sequence again.
3. Repeat the above steps till angles from 0 to 180° are covered.

Step 1 is the time sequence of projection data for a given view angle. The entire experimental data set comprises the time sequences for all view angles. In schlieren imaging, light intensity (specifically, the contrast) scales linearly with the density gradient. Hence, numerical calculations can be performed directly with light intensity. Reconstructions are performed plane-by-plane within the experimental apparatus. The modes of the time sequence of light intensity values along a row of the schlieren image constitute the data set for tomography.

The experimental data is now reduced as per the following algorithm:

1. Start with the image data for a given view angle.
2. Select a horizontal plane above the growing crystal from the schlieren image where the reconstruction is to be performed.
3. For the selected plane, form the POD data matrix. For a given time instant, express the intensity values as the column of the matrix. Similarly intensity data for other time instants form the remaining columns.
4. Subtract out the average of every column; modal analysis is performed for the mean removed data.
5. For the rectangular matrix thus obtained, determine the POD basis vectors and the corresponding time components.
6. Repeat steps 1–5 for all view angles. Modes of each projection (between 0 and 180°) have now been obtained by considering all the time instants.
7. Use a tomography algorithm (direct such as CBP, or iterative) to convert the modes of projection data into the modes of concentration gradient over selected horizontal plane above the crystal.
8. Repeat step 7 for as many modes as are significant.
9. Determine the reconstructed time-dependent concentration gradient field using the time components available at step 5.

For validation of the combined POD-tomographic algorithm against simulated as well as experimental data, the reader is referred to [29].

References

1. Anderson AH, Kak AC (1984) Simultaneous algebraic reconstruction technique (SART): a superior implementation of the ART algorithm. Ultrason Imaging 6:81–94
2. Bahl S, Liburdy JA (1991) Three dimensional image reconstruction using interferometric data from a limited field of view with noise. Appl Opt 30(29):4218–4226
3. Bahl S, Liburdy JA (1991) Measurement of local convective heat transfer coefficients using three-dimensional interferometry. Int J Heat Mass Transf 34:949–960
4. Censor Y (1983) Finite series-expansion reconstruction methods. Proc IEEE 71(3):409–419
5. Faris GW, Byer RL (1988) Three dimensional beam deflection optical tomography of a supersonic jet. Appl Opt 27(24):5202–5212
6. Gilbert PFC (1972) Iterative methods for three-dimensional reconstruction of an object from its projections. J Theor Biol 36:105–117
7. Gordon R, Bender R, Herman GT (1970) Algebraic reconstruction techniques (ART) for three-dimensional electron microscopy and X-ray photography. J Theor Biol 29:471–481
8. Gordon R, Herman GT (1974) Three dimensional reconstructions from projections: a review of algorithms. Int Rev Crystallogr 38:111–151
9. Gull SF, Newton TJ (1986) Maximum entropy tomography. Appl Opt 25:156–160
10. Herman GT (1980) Image reconstruction from projections. Academic Press, New York
11. Kaczmarz MS (1937) Angenaherte auflosung von systemen linearer gleichungen. Bull Acad Polonaise Sci Lett Classe Sci Math Natur Serier A35:355–357
12. Lanen TAWM (1990) Digital holographic interferometry in flow research. Opt Commn 79:386–396
13. Liu TC, Merzkirch W, Oberste-Lehn K (1989) Optical tomography applied to speckle photographic measurement of asymmetric flows with variable density. Exp Fluids 7:157–163
14. Mayinger F (eds) (1994) Optical measurements: techniques and applications. Springer, Berlin
15. Mewes D, Friedrich M, Haarde W, Ostendorf W (1990) Tomographic measurement techniques for process engineering studies. In: Cheremisinoff NP (ed) Handbook of heat and mass transfer, Chapter 24, vol 3
16. Michael YC, Yang KT (1992) Three-dimensional mach-zehnder interferometric tomography of the rayleigh-benard problem. J Heat Transf Trans ASME 114:622–629
17. Mishra D, Muralidhar K, Munshi P (1998) Performance evaluation of fringe thinning algorithms for interferometric tomography. Opt Lasers Eng 30:229–249
18. Mishra D, Muralidhar K, Munshi P (1999a) Interferometric study of rayleigh-benard convection using tomography with limited projection data. Exp Heat Transf 12(2):117–136
19. Mishra D, Muralidhar K, Munshi P (1999c) A robust MART algorithm for tomographic applications. Numer Heat Transf B Fundam. 35(4):485–506
20. Mishra D, Muralidhar K, Munshi P (1999d) Interferometric study of rayleigh-benard convection at intermediate rayleigh numbers. Fluid Dynamics Res 25(5):231–255
21. Mukutmoni D, Yang KT (1995) Pattern selection for rayleigh-benard convection in intermediate aspect ratio boxes. Numer Heat Transf Part A 27:621–637
22. Munshi P (1997) Application of computerized tomography for measurements in heat and mass transfer, proceedings of the 3rd ISHMT-ASME heat and mass transfer conference held at IIT Kanpur (India) during 29–31 December 1997. Narosa Publishers, New Delhi
23. Natterer F (1986) The mathematics of computerized tomography. Wiley, New York
24. Ostendorf W, Mayinger F, Mewes D 1986 A tomographical method using holographic interferometry for the registration of three dimeinsonal unsteady temperature profiles in laminar and turbulent flow, proceedings of the 8th international heat transfer conference, San Francisco, USA, pp 519–523
25. Sirovich L (1989) Chaotic dynamics of coherent structures. Physica D 37:126–145
26. Snyder R, Hesselink L (1985) High speed optical tomography for flow visualization. Appl Opt 24:23
27. Snyder R (1988) Instantaneous three dimensional optical tomographic measurements of species concentration in a co-flowing jet, Report No. SUDAAR 567, Stanford University, USA

28. Soller C, Wenskus R, Middendorf P, Meier GEA, Obermeier F (1994) Interferometric tomography for flow visualization of density fields in supersonic jets and convective flow. Appl Opt 33(14):2921–2932
29. Srivastava A, Singh D, Muralidhar K (2009) Reconstruction of time-dependent concentration gradients around a KDP crystal growing from its aqueous solution. J Crystal Growth 311:1166–1177
30. Subbarao PMV, Munshi P, Muralidhar K (1997) Performance evaluation of iterative tomographic algorithms applied to reconstruction of a three dimensional temperature field. Numer Heat Transf B Fundam 31(3):347–372
31. Tanabe K (1971) Projection method for solving a singular system. Numer Math 17:302–214
32. Tolpadi AK, Kuehn TH (1991) Measurement of three-dimensional temperature fields in conjugate conduction-convection problems using multidimensional interferometry. Int J Heat Mass Transfer 34(7):1733–1745
33. Torniainen ED, Hinz A, Gouldin FC (1998) Tomographic analysis of unsteady. Reacting Flows AIAA J 36:1270–1278
34. Velarde MG, Normand C (1980) Convection. Sci American 243(1):79–94
35. Verhoeven D (1993) Multiplicative algebraic computed tomography algorithms for the reconstruction of multidirectional interferometric data. Opt Eng 32:410–419
36. Watt DW, Vest CM (1990) Turbulent flow visualization by interferometric integral imaging and computed tomography. Exp Fluids 8:301–311

Chapter 5
Validation Studies

Keywords Mach–Zehnder interferometer · Hotwire anemometer · Cylinder wake · Crystal growth · Buoyant jets

5.1 Introduction

Optical imaging is a useful tool for flow visualization. When used as a quantitative technique in thermal measurements, the procedure needs to be carefully benchmarked and validated. These comparisons can be against independent measurements and against published data. Over and above, the images should satisfy certain internal consistency checks. These include checks on mass and energy balance on various scales. When the field being measured is unsteady, it may be difficult to compare instantaneous data across various sources. Under these conditions, the comparison is in terms of the statistics of the time-dependent flow property. In the present chapter, schlieren and shadowgraphy techniques are evaluated for quantitative characterization in various flow configurations. The comparison is against independent measurements using a Mach–Zehnder interferometer and a hotwire anemometer. In addition, comparison is reported against published data. Configurations considered for comparison are the following:

1. Convection in a differentially heated fluid layer,
2. Convection patterns in a crystal growth process,
3. Two-layer convection,
4. Buoyant jets, and
5. Wake of a heated cylinder.

P. K. Panigrahi and K. Muralidhar, *Schlieren and Shadowgraph Methods in Heat and Mass Transfer*, SpringerBriefs in Thermal Engineering and Applied Science, DOI: 10.1007/978-1-4614-4535-7_5, © The Author(s) 2012

5.2 Convection in a Differentially Heated Fluid Layer

The apparatus used to study the buoyancy-driven flow in a horizontal fluid layer is shown schematically in Fig. 5.1. The cavity is 447 mm long with a square cross-section of edge 32 mm. The test cell consists of three sections, namely, the top plate, the fluid layer enclosed in a cavity, and the bottom plate. The top and bottom walls of the cavity are made from 3 mm thick aluminum plates. The flatness of these plates was manufacturer-specified to be within ±0.1 mm and was further improved during the fabrication of the apparatus. The central portion of the experimental apparatus is the test section containing the fluid medium. The sidewalls of the cavity were made of a 10 mm plexiglas sheet. In turn, the plexiglas sheet was covered with a thick bakelite padding in order to insulate the test section with respect to the atmosphere. The height of the test section was 32 mm and was measured to be uniform to within ±0.1 mm. Optical window was provided in the direction of propagation of the laser beam. It was held parallel to the longest dimension of the cavity for recording the projected convective field in the form of 2D images. The apparatus was enclosed in a larger chamber made of thermocole to minimize the influence of external temperature variations. The room temperature during the experiments was a constant to better than ±0.5 °C over a 10–12 h period. The thermal fields in air stabilized over a period of 5–6 h: In water, dynamically steady patterns were realized in 1–2 h. In the experiments, the thermally active surfaces were maintained at uniform temperatures by circulating a large volume of water over them from constant temperature baths. Temperature control of the baths was rated as ±0.1 °C at the cavity location, direct measurements with a multi-channel temperature recorder showed a spatial variation of less than ±0.2 °C. For the upper plate, a tank-like construction enabled extended contact between the flowing water and the aluminum surface. Special arrangements are required to maintain good contact between water and the lower surface of the plate. Aluminum baffles introduced a tortuous path to flow, serve this purpose by increasing the effective interfacial contact area. The passage of the light beam was along the longer edge of the cavity. The respective ends were closed using optical windows.

The temperature difference in the vertical direction leads to unstable density gradients in the fluid medium. Hence, a buoyancy-driven convection pattern is setup in the cavity. Transient as well as steady state thermal patterns have been recorded. The temperature distribution in the stratified fluid layer and wall heat fluxes are presented and compared with the literature, where available.

5.2.1 Nusselt Number Calculation

Heat transfer rates at the heated and cooled walls of the cavity are expressed in terms of the Nusselt number defined as follows:

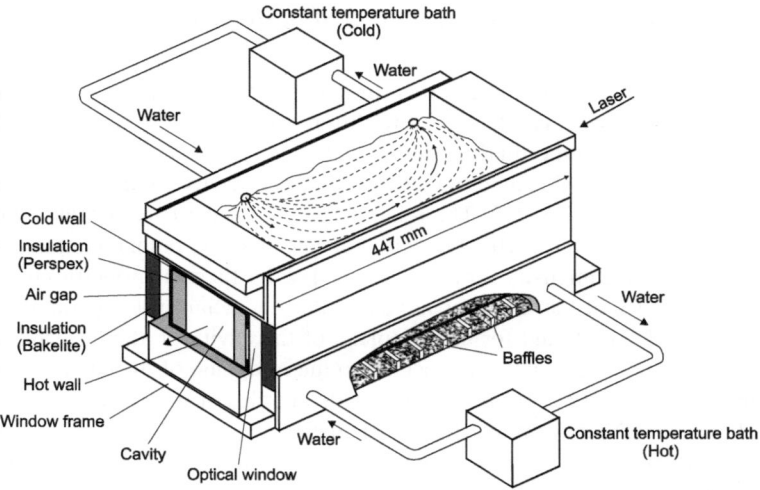

Fig. 5.1 Experimental apparatus for the study of Rayleigh–Benard convection in a rectangular cavity

$$\mathrm{Nu} = -\frac{H}{T_{\mathrm{hot}} - T_{\mathrm{cold}}} \left. \frac{\partial T}{\partial y} \right|_{y=0,H} \tag{5.1}$$

Here H is the height of the fluid layer and y is a coordinate in the vertical direction. The average Nusselt number is obtained by integrating the above expression over the width of the cavity. Since temperature in Eq. 5.1 is an average in the viewing direction of the laser beam, the local Nusselt number should be interpreted in a similar manner. The average Nusselt number for each of the horizontal surfaces has been compared with the experimental correlation reported by Gebhart et al. [2]. In air, this correlation is given as:

$$\mathrm{Nu(air)} = 1 + 1.44 \left[1 - \frac{1708}{\mathrm{Ra}} \right] + \left[\left(\frac{\mathrm{Ra}}{5830} \right)^{\frac{1}{3}} - 1 \right], \quad \mathrm{Ra} < 10^6 \tag{5.2}$$

With ν and α representing the kinematic viscosity and thermal diffusivity, respectively, Rayleigh number is defined as

$$\mathrm{Ra} = \frac{g\beta(T_{\mathrm{hot}} - T_{\mathrm{cold}})H^3}{\nu\alpha} \tag{5.3}$$

5.2.2 Interferometry, Schlieren, and Shadowgraph

A direct comparison of the images seen in interferometry, schlieren, and shadow-graph is presented here. The temperature difference across the cavity was 10 K in the experiments, while the fluid medium in the cavity was air. The corresponding Rayleigh number was calculated to be 6×10^4. The respective images are compared on the right column of Fig. 5.2. The relative spacing of the fringes yields the temperature profile in interferometry. For schlieren and shadowgraph, the information is present in the relative intensity variation. The respective thermal properties recovered are the local values of the first derivative and the Laplacian of temperature. These quantities have been plotted for the midplane of the cavity on the left side of Fig. 5.2. The individual data points are specific to the midplane of the cavity, while the solid line indicates the overall trend. The shaded circles of the left column indicate the gradients calculated from interferometric data (for schlieren) and from schlieren data (for shadowgraph). Overall, a good match is a confirmation of the result that schlieren is a derivative of the interferometric field, and the shadowgraph, to a first approximation, is a derivative of the schlieren. The appearance of dense fringes near the horizontal walls is indicative of high temperature gradients at these locations. This is brought in the schlieren image, as well as the data points. The central region is a zone of nearly constant temperature, where the gradients are close to zero. Thus, the schlieren images and interferograms correlate quite well with each other. These also correlate with the shadowgraph, once it is realized that in this approach, light is redistributed over the image. In a shadowgraph image, light from the region close to cold top wall deflects toward the lower hot wall, where the light intensity shows a maximum. Thus, high gradients are represented in a shadowgraph by regions of very low as well as very high light intensity. In the central core, the change in light intensity with respect to the initial setting is small. Thus the Laplacian of temperature in this region is practically zero. The thermal lensing effect that distorts the shape of the cavity cross-section is most visible in the shadowgraph.

5.2.3 Rainbow Schlieren Technique

For validation of the rainbow (color) schlieren technique (Chap. 3), buoyancy-driven convection in an air-filled rectangular cavity is considered once again. The transient evolution of the flow field was recorded at a regular time intervals till practically steady state was reached. Figure 5.3 shows the transient evolution of the thermal field in air in the form of color images.

Initially, temperature gradients are absent and the light beam is focused at the orange region of the filter. Figure 5.3a shows the base image of the experiment. A temperature difference of 10 K (Ra = 3.2×10^4) is applied between the upper and lower surfaces of the cavity by circulating cold and hot water in the respective tanks. The temperature difference across the cavity leads to an unstable density gradient in

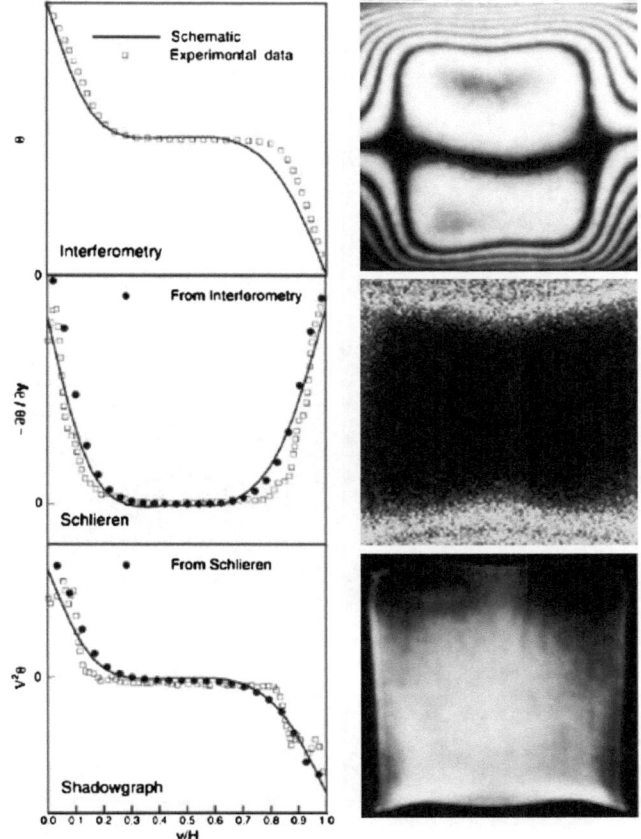

Fig. 5.2 Comparison of data recovered from the three optical techniques (*left column*). The corresponding experimental images are shown in the *right column*. Ra $= 6 \times 10^4$. The *solid line* in the *left column* is representative of the trend seen at any vertical section of the image

the fluid medium. Hence, convection currents are set up in the cavity. At this stage, the light beam deflects from its original path due to changes in the index of refraction of the medium and falls on a new range of colors, Fig. 5.3b and c. With the passage of time, the flow patterns approach steady state. For the present experiment, steady state was achieved in 3 h.

Figure 5.3d shows the steady state patterns recorded by the color schlieren technique. As expected, changes in color are prominent near the heated and cooled surfaces. The image shows rapid color variation near the horizontal walls while the central region is one of uniform color, close to the base image. This is characteristic of cellular convection formed in a rectangular cavity. Since the sidewalls are thermally insulated, gradients within the fluid are visible but the gradient at the wall

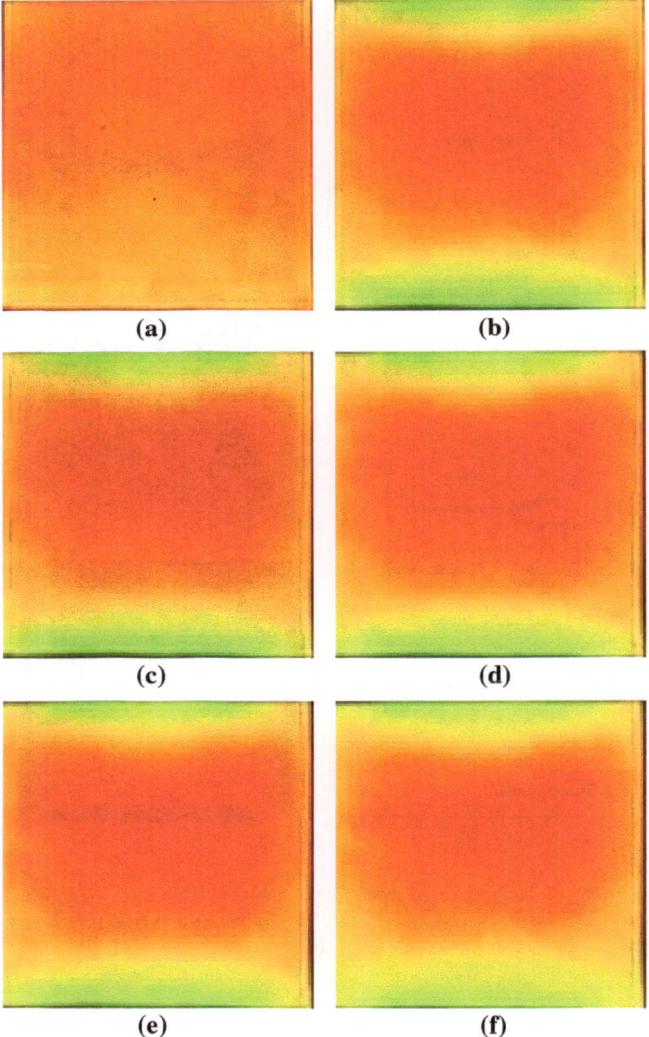

Fig. 5.3 Evolution of convective field in the air-filled rectangular cavity at ΔT=10 K and Ra = 32, 000

itself is zero. The convection pattern is nearly symmetric about the centerline of the cavity and is realized in the color schlieren image.

During data analysis in color schlieren, the information is extracted in terms of the hue distribution. Figure 5.4a shows the hue variation with respect to the non-dimensionlized vertical coordinate for $x/H = 1/4, 1/2$ and 3/4. It is seen that the hue variation is large near the walls and nearly constant within $0.2 < y/H < 0.8$.

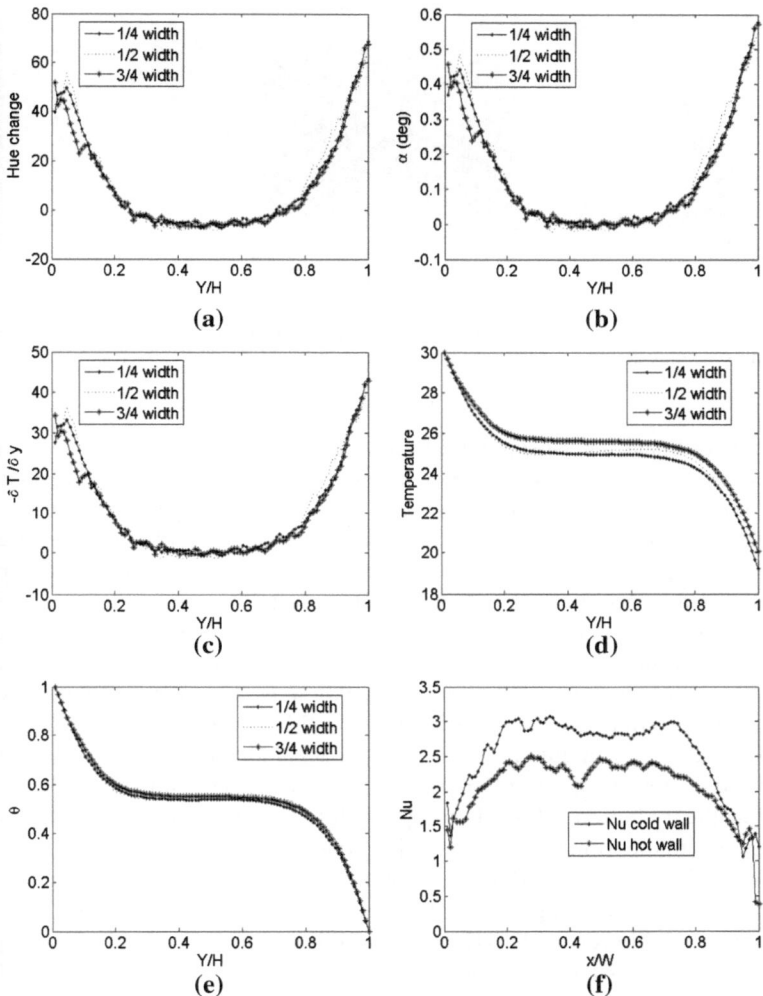

Fig. 5.4 Variation of **a** hue, **b** beam deflection (α), and **c** temperature gradient with respect to the non-dimensional vertical coordinate. **d** Temperature profiles evaluated with temperature of the top wall and that of the lower wall used independently as boundary conditions. **e** Non-dimensional temperature profiles at various columns within the cavity. **f** Local Nusselt number distribution at the hot and cold walls of the rectangular cavity for $\Delta T = 10\,\mathrm{K}$ and $\mathrm{Ra} = 32,000$

Figure 5.4b shows the beam deflection (α) in degrees with respect to the vertical coordinate (y/H). It shows that the deflection of light is high near the walls though it deflects very little in the central region of the cavity. Equivalently, the refractive index variation is high near the walls and small in the central portion of the cavity. The thermal property recovered from beam deflection is the first derivative of temperature. Figure 5.4c shows the variation of temperature gradient with respect

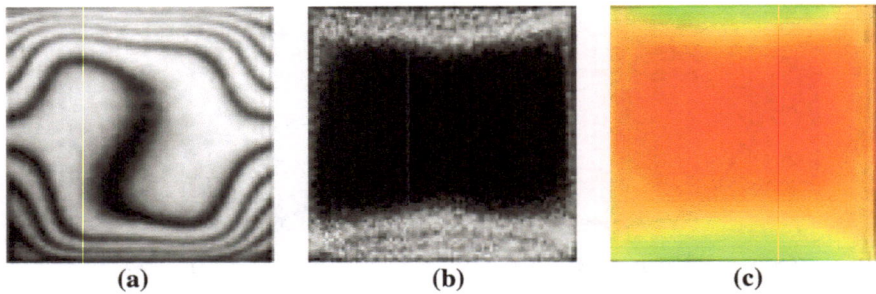

Fig. 5.5 **a** M–Z interferogram, **b** monochrome schlieren image, and **c** color schlieren image for the steady state convective flow patterns in air-filled cavity at $\Delta T = 10$ K for Ra $= 32,000$

to the y coordinate for various x-locations. The central region is a zone of nearly constant temperature, where the gradients are close to zero. Figure 5.4d shows the temperature profiles evaluated across the cavity from the gradients. One profile uses the boundary condition at the top wall to start integration while the other uses the lower wall temperature. The profiles obtained are seen to match well. The data points in this figure are specific to the midplane of the cavity.

Non-dimensional temperature profiles at three columns ($x/h = 1/4, 1/2$ and $3/4$) are shown in Fig. 5.4e. The shape of the profile, characteristic of buoyancy-driven convection is realized at all the three columns. The slopes of the individual curves near the walls give a measure of the wall heat flux. Figure 5.4f shows the dimensionless wall heat flux in the form of the local Nusselt number distribution at the top and the lower walls ($y/H = 0$ and 1). Larger Nusselt numbers near the center arise from the formation of boundary-layers while near the corners, Nusselt numbers decrease due to the formation of stagnation zones.

Figure 5.5 shows an interferogram, monochrome schlieren image, and a color schlieren image for Rayleigh–Benard convection in an air-filled rectangular cavity for temperature difference of 10 K corresponding to Ra $= 3.2 \times 10^4$. The interferogram directly gives the temperature field in the fluid medium whereas schlieren images provide temperature gradients. A comparison of the local steady dimensionless temperature profiles in the cavity as obtained from M–Z interferometry, monochrome schlieren, and color schlieren is presented in Fig. 5.6. The comparison among the experiments is seen to be good.

The average Nusselt number for each of the surfaces has also been compared with the experimental correlation reported by Gebhart et al. [2]. The wall averaged Nusselt numbers calculated from the color schlieren images by using Eq. 5.1 is 3.14 for the cold wall and 3.24 for the hot wall. The Nusselt number calculated from the correlation, Eq. 5.2, is 3.13 with a stated uncertainty of ± 20%. The large uncertainty arises from the fact that the convection pattern is 3D and heat transfer depends on the cavity aspect ratio.

Figure 5.7 shows color schlieren images that map the evolution of convection in an air-filled rectangular cavity for a cavity temperature difference of 15 K. These

Fig. 5.6 Comparison of temperature profiles obtained from color schlieren technique with M–Z interferometry and monochrome schlieren techniques

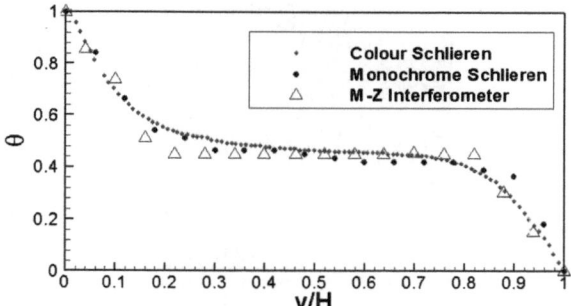

patterns are quite similar to the experiment wherein $\Delta T = 10\,\mathrm{K}$. Figure 5.7a shows the base image and Fig. 5.7b shows the convection pattern at steady state, specifically after the passage of 6 hours. It is to be expected that temperature gradients, and hence the range of colors would be higher for 15 K experiment. This trend is confirmed from the steady state image of 10 K, Fig. 5.3f to be compared with Fig. 5.7b for a cavity temperature difference of 15 K. Figure 5.8a shows the hue variation with respect to the non-dimensional vertical coordinate for $x/h = 1/4, 1/2$ and 3/4. Figure 5.8b shows the beam deflection (α) in degrees with respect to the vertical coordinate (y/H). Once again, the deflection of light is seen to be quite high near the walls and very little at the center. Figure 5.8c shows the variation of temperature gradient with respect to the y coordinate for various x-locations. The central region is a zone of nearly constant temperature, where the gradients are close to zero. Figure 5.8d shows the temperature profiles evaluated across the cavity, once with the boundary condition at the top wall and next, the lower wall. The profiles match quite well in the near wall regions while a small deviation is observed in the central part of the cavity. Non-dimensional temperature profiles at three columns $(x/h = 1/4, 1/2$ and 3/4) are shown in Fig. 5.8e. The shape of the profile is characteristic of buoyancy-driven convection.

Figure 5.8f shows the local Nusselt number distribution at the top wall $(y/H = 1)$ and the lower wall $(y/H = 0)$. The Nusselt number peaks at the cold wall are slightly larger than the hot wall, but the overall variations are quite similar. Under steady state conditions, the average wall heat flux is also equal to the energy transferred across any horizontal plane of the cavity and is a constant, whether calculated at the hot or the cold wall. From Fig. 5.8f the wall-averaged Nusselt number calculated from the color schlieren image is 3.38 for the cold wall and 3.72 for the hot wall. The Nusselt number calculated from the correlation of Eq. 5.2 is 3.41 ($\pm 20\,\%$). The three values are quite close to each other. Differences arise from intrinsic flow unsteadiness as well as assumptions of small angle deflections during data analysis.

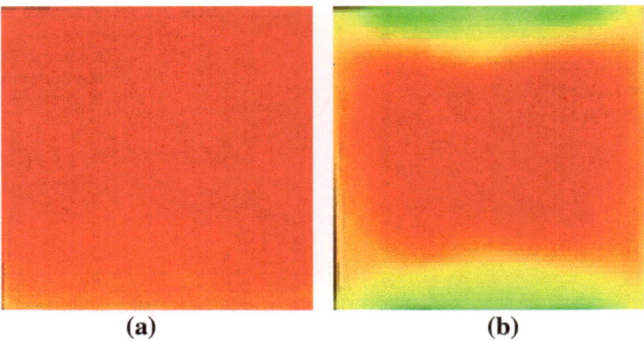

(a)   (b)

Fig. 5.7 Color schlieren: evolution of the buoyancy-driven convective field in an air-filled rectangular cavity for $\Delta T = 15\,\text{K}$ and Ra $= 48,000$. **a** is the base image while **b** indicates long-term convection pattern

5.3 Convection Patterns in a Crystal Growth Process

Imaging convection patterns around a crystal growing from its aqueous solution is discussed; for an introduction, see [6, 7]. Comparison of color and monochrome schlieren images of convection around a potassium dihydrogen phosphate (KDP) crystal growing from its aqueous solution is reported in the present section. A crystal growth chamber shown in Fig. 5.9 is used for KDP crystal growth. It has two parts: the inner crystal growth cell and the outer water tank carrying flow of thermostated water. The inner cell is a glass beaker whereas the outer chamber is made of Plexiglas. At the opposite ends on the beaker walls, two circular optical windows (BK-7 glass, 60 mm diameter) are mounted for optical access of the crystal growth process. Supersaturated solution of KDP in water fills the inner beaker and a seed crystal is introduced inside the growth chamber. For the preparation of supersaturated solution at a particular temperature, the amount of solute to be dissolved is determined from the solubility curve available in the literature. The supersaturated solution is prepared at a temperature of 37 °C.

5.3.1 Convective Patterns

The KDP crystal growth process from an aqueous solution relies on the deposition of the excess salt in the supersaturated solution at the surface of the growing crystal. The deposition rate, in turn, depends on the size of the crystal, the initial salt concentration, and the cooling rate. The crystal grows a few mm over a period of a few days. The mass flux of salt depositing on the crystal surface depends on convection patterns arising from density (and hence concentration) differences within the solution. Convection currents, essentially, bring fresh solute to the vicinity of the seed.

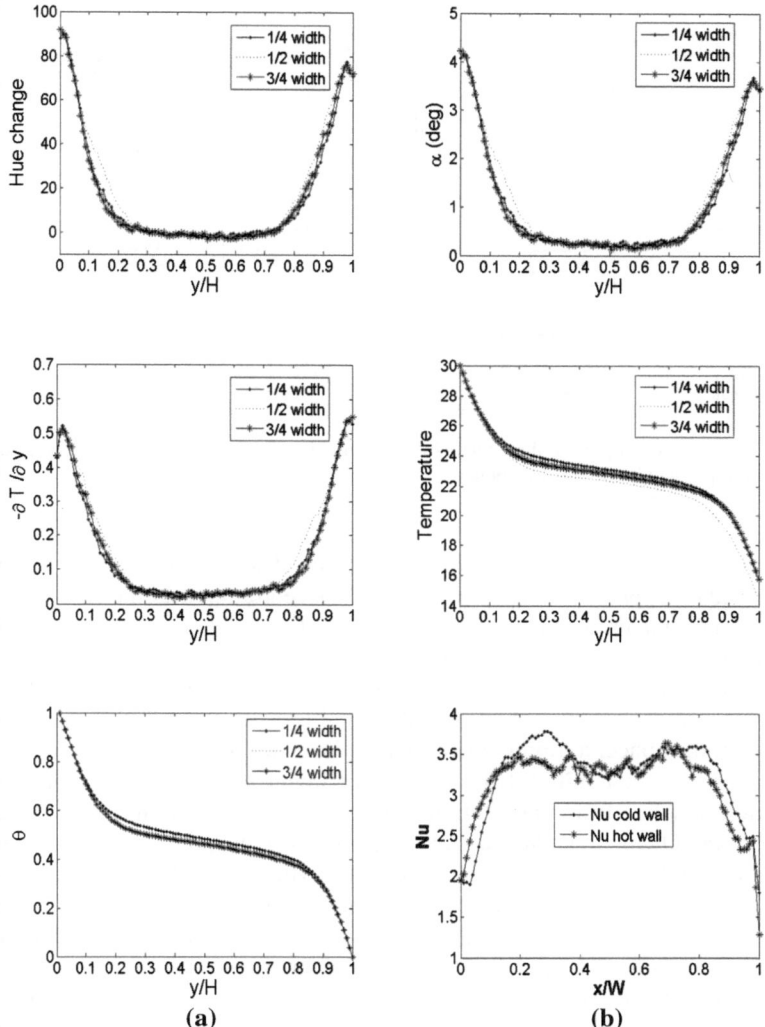

Fig. 5.8 Variation of **a** hue, **b** beam deflection (α), and **c** temperature gradient with respect to the non-dimensional vertical coordinate. **d** Temperature profiles evaluated with temperature of the top wall and that of the lower wall used independently as boundary conditions. **e** Non-dimensional temperature profiles at various columns within the cavity. **f** Local Nusselt number distribution at the hot and cold walls of the rectangular cavity for $\Delta T = 15\,\text{K}$ and $\text{Ra} = 48{,}000$

The time sequence of the convection patterns that arise during the growth of the KDP crystal is shown in Fig. 5.10. Fluid convection is related to the fact that solution adjacent to the crystal is depleted of salt and is lighter than the rest of the medium. In Fig. 5.10, the seed holder is a glass rod, kept vertical, and appears black; at the

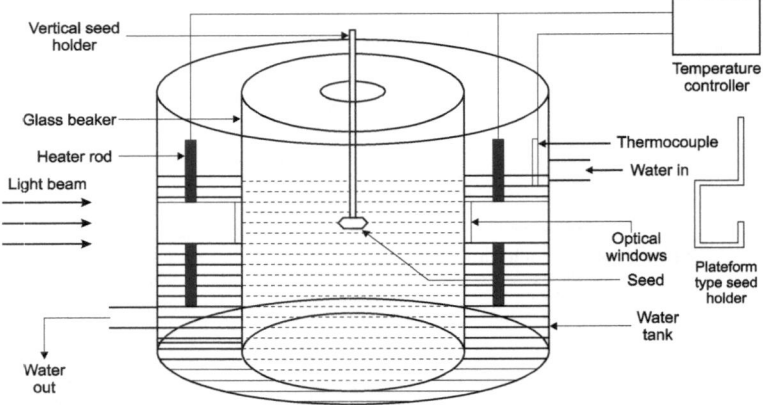

Fig. 5.9 Schematic diagram of the KDP crystal growth chamber with hanging seed configuration; a platform type seed holder is also shown

lower end of the seed holder is the growing crystal. The deposition of the salt on the seed crystal takes place leading to a change in concentration near the crystal surface. Changes in the material density (and hence, refractive index) displace the light beam and a color distribution is obtained on the filter plane.

The schlieren images in Fig. 5.10 show plumes of lower color contrast for the smaller crystal during the early stage of crystal growth. Since the change in color scales with the concentration gradient (Chap. 3), it can be inferred that density gradients are small in the initial stages of the experiment. With the passage of time, the deposition of solute from the solution takes place leading to an increase in the size of the crystal. Owing to an increase in the crystal size, the plume strength increases. The plume width also increases when compared to the initial. As the crystal grows in size, convection intensifies, and the growth rate is enhanced. The process continues till the solution around the crystal is deprived of excess salt. For the duration of the experiment, Fig. 5.10 shows that the plumes are quasi-steady on the short time scale but respond to the long-time changes in the crystal size and supersaturation. The plumes are symmetric and originate from the crystal edge where the concentration gradient is possibly the greatest. The color variation outside the plume is small and one can conclude that salt concentration gradients are small in the bulk of the solution.

The color schlieren technique is sensitive enough to unwanted factors present during the crystal growth process. For example, Fig. 5.10d shows two plumes emerging below the crystal related to unwanted nucleation at the base of the growth chamber.

Figure 5.11 shows images of a crystal growth experiment conducted in a platform arrangement. In this experiment the seed is placed over a rod supported over a platform. Convection plumes above the crystal are seen once again. Above the crystal, layers of solution with high concentration gradually move down to replace the less dense solution. Since the crystal is grown from a fixed volume of solution,

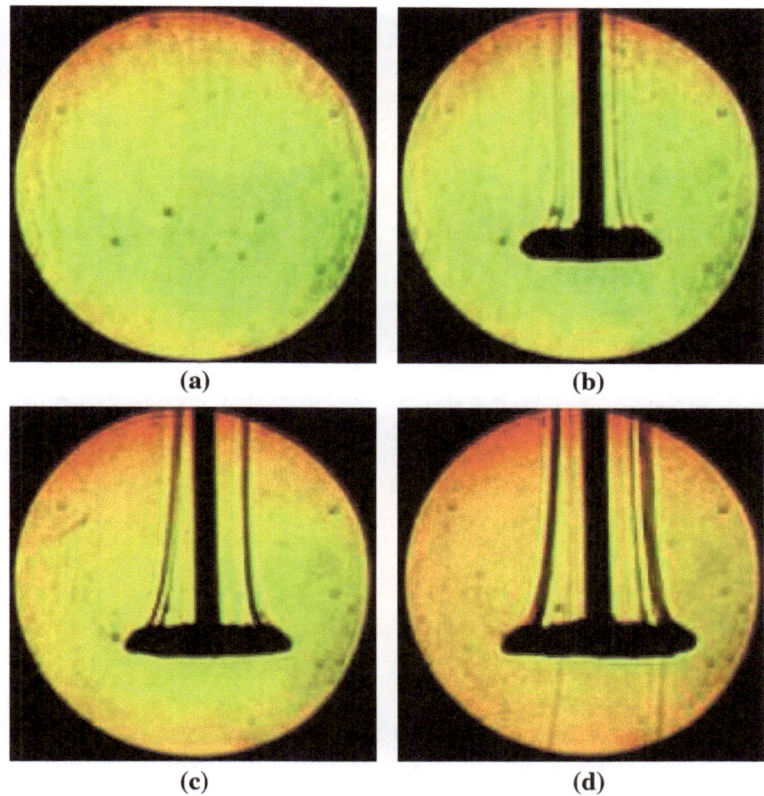

(a)   (b)

(c)   (d)

Fig. 5.10 Rainbow schlieren imaging of evolution of the convective field during KDP crystal growth. **a** Base image. **b** $t = 1$ h. **c** $t = 15$ h. **d** $t = 20$ h

the growth chamber is depleted of salt and develops density stratification. After 40 h, stratification is distinctly seen below the crystal. It is characterized by layering of the solution in a gravitationally stable configuration. Density inversion suppresses convection to a point where further increase in the crystal size is negligible.

In the experiment being discussed, it is interesting to see two stratification zones, one below the crystal and one above—seen as a band of faint indigo. To ensure recording the two stratification zones, images are recorded by moving the color filter. Figure 5.12 shows the stratification in the solution after 40 h in three different color zones of the filter. From Fig. 5.12a, b two bands are clearly seen but Fig. 5.12c does not show the stratification band. In this respect, the color filter provides an additional degree of freedom for selective imaging of the concentration field.

The color schlieren image (Fig. 5.13d) of the KDP crystal growth process is compared with those of the Mach–Zehnder interferometer (Fig. 5.13a), monochrome schlieren (Fig. 5.13b), and shadowgraph (Fig. 5.13c). Interferograms are recorded in the wedge fringe setting. An abrupt change in solute concentration around the

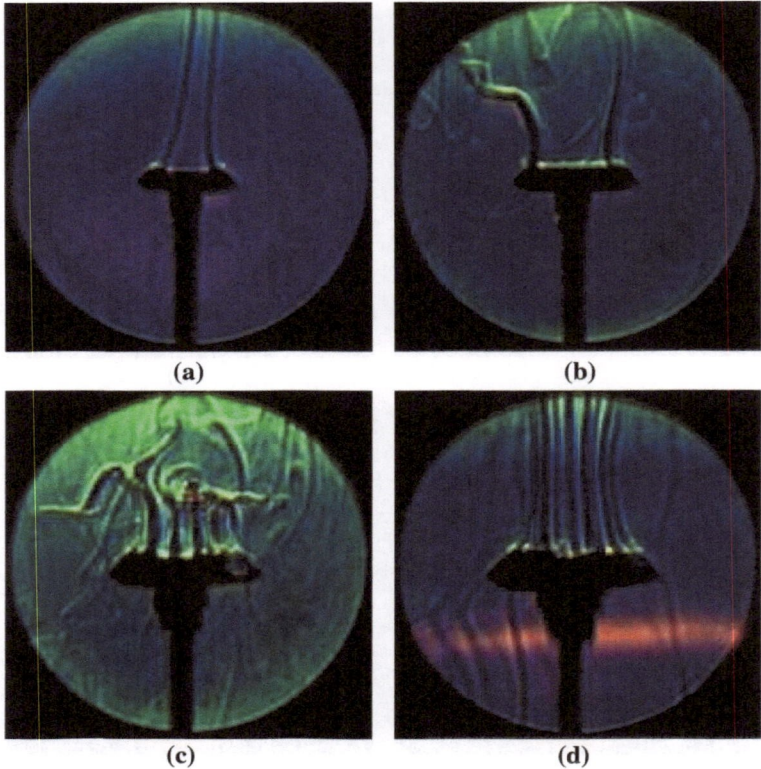

Fig. 5.11 Evolution of the convective field during KDP crystal growth in a platform arrangement. A gradual increase in the size of the crystal is visible. **a** After 15 min. **b** $t = 1$ h. **c** $t = 30$ h. **d** $t = 40$ h

seed crystal is seen in terms of fringe displacement in interferometry and intensity contrast in the schlieren and shadowgraph images. The concentration boundary layer creates a jump in the refractive index, which deflects the light beam into the region of relatively higher salt concentration. The monochrome schlieren image shows better contrast than the shadowgraph.

Figure 5.14 compares the concentration profile from color schlieren with that obtained from monochrome schlieren at a selected time instant in two separate experiments. Concentration is normalized with the value prevailing in the bulk of the solution. A concentration value of '0' represents the saturated state and '1' represents the supersaturated condition at the temperature of the solution in the growth chamber. The concentration distributions recorded by two different techniques are quite similar. Differences can be attributed to slight differences in the cooling rate, the crystal size and the time elapsed. The quantitative utility of the schlieren approach is, however, clearly brought out.

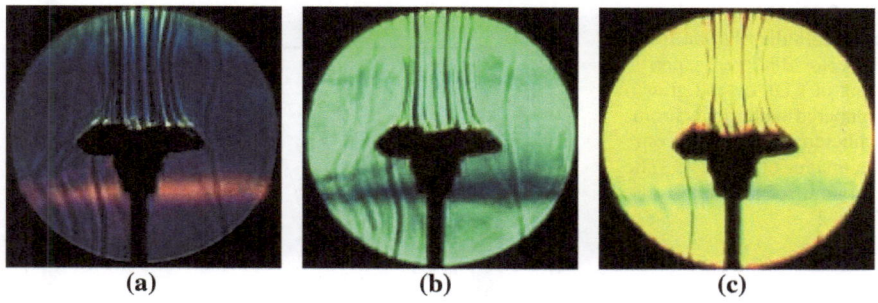

Fig. 5.12 Stratification within the solution visualized at three color locations of filter

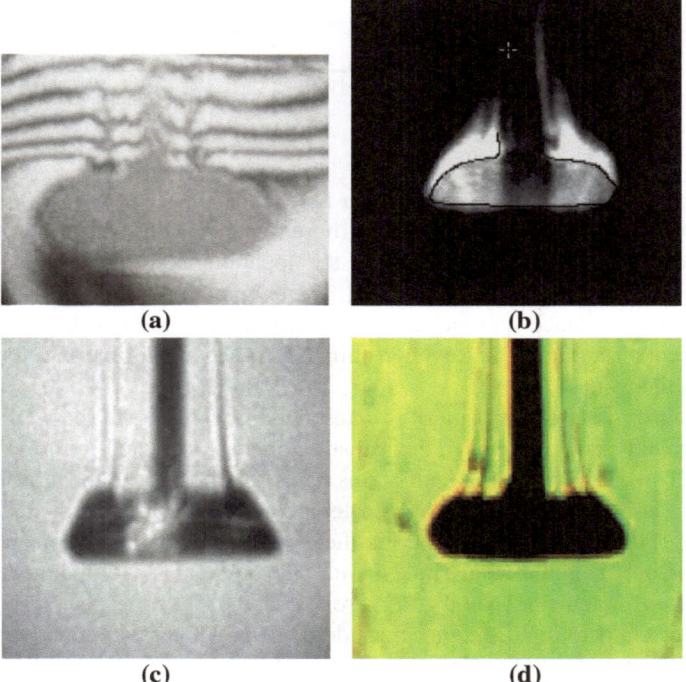

Fig. 5.13 Comparison among **a** Mach–Zehnder interferogram, **b** monochrome schlieren image, **c** shadowgraph and **d** color schlieren image recorded in a KDP crystal growth experiment

5.4 Steady Two-Layer Convection

Experiments in steady two-layer convection are reported in the present section; see [3] for an introduction to the subject. The fluid combination considered is air and silicone oil in a cavity that is octagonal in plan. The cavity is an approximation of a

Fig. 5.14 Normalized concentration distribution from
the color schlieren experiment of KDP crystal growth
compared with monochrome
schlieren; data drawn from
the images of Fig. 5.13. The
literature referred is [9] of
Chap. 2

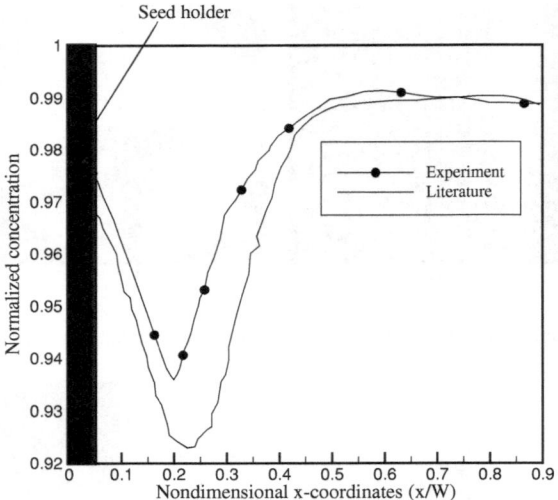

Fig. 5.14 Normalized concentration distribution from the color schlieren experiment of KDP crystal growth compared with monochrome schlieren; data drawn from the images of Fig. 5.13. The literature referred is [9] of Chap. 2

circular axisymmetric geometry. Oil layer heights of 30, 50, and 70 % of the cavity height have been studied. Temperature differences (ΔT) of 0.4, 2.4, and 5 K were applied across the cavity. The details of the apparatus and the complete range of parameters considered are provided in [5]. The temperature at the interface obtained from the energy balance approach is abbreviated as *estimated interface temperature* in the present discussion. The Rayleigh number is based on this interface temperature relative to the walls.

The planform of cellular pattern is determined largely by the shape of the apparatus. In an axisymmetric cavity, the roll structures are expected to form concentric rings. The interference fringe field forms as a superposition of several rolls. The fringes are contours of depth-averaged temperature and arrange themselves in a symmetric (Ω) pattern. The full thermal field is a collection of several such omega and inverse omega patterns interlinked with one another. The temperature drop per fringe shift in air and silicone oil (ΔT_ϵ) are 5.65 and 0.012 K, respectively. Thus, for the range of temperature difference considered here, fringes were not seen in air. In silicone oil, a dense set of fringes were obtained.

For overall temperature differences of 0.4 and 2.8 K, experimental data are presented in the form of interferograms. For $\Delta T = 5$ K, refraction errors in oil are large and interferometric analysis is not possible. For this experiment, shadowgraphs have been evaluated. Data reduction using the two imaging techniques are discussed in the following section.

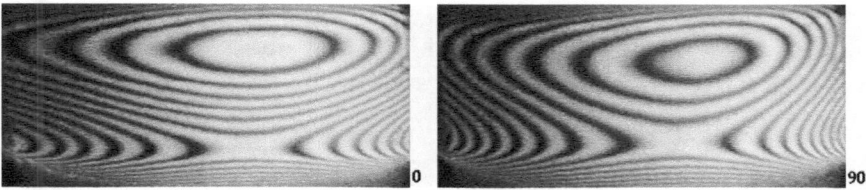

Fig. 5.15 Interferograms corresponding to $\Delta T = 0.4$ K for view angles of $0°$ (*left*) and $90°$ (*right*) in a cavity 50 % filled with silicone oil

Fig. 5.16 Line-of-sight averaged temperature profile in silicone oil for $\Delta T = 0.4$ K for the two projection angles of Fig. 5.15

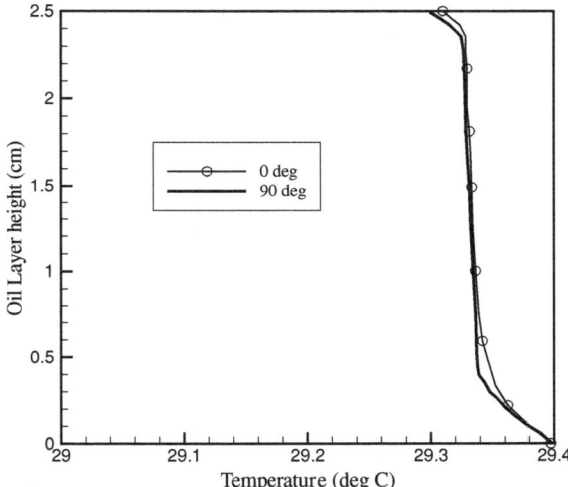

5.4.1 Temperature Variation in Silicone Oil

The hot and cold wall temperatures in the experiment were 29.4 and 29.0 °C, respectively. The oil layer height is 2.5 cm, which is half the cavity height of 5 cm. The interface temperature determined from fringe counting was obtained as 29.30 °C. The energy balance approach of [5] yielded a value of 29.33 °C. Rayleigh number based on the interface temperature was 607 in air and 2072 for silicone oil. Thus, one can expect weak, steady, axisymmetric convection in silicone oil, while air would be in the conduction regime. Figure 5.15 shows interferograms recorded in oil for two view angles, namely 0 and 90 °. The fringe pattern indicates the thermal field to be axisymmetric. The line-of-sight averaged temperature distribution in the vertical direction for 0 and 90 °, shown in Fig. 5.16, are understandably very close.

Figure 5.17 shows interferograms for a temperature difference of 2.4 °C across the cavity. The hot wall temperature was 29.4 °C while the cold wall was changed to 27.0 °C. The experimental and estimated interface temperatures were 29.20 °C and 29.11 °C. The Rayleigh numbers in air and silicone oil were determined with respect

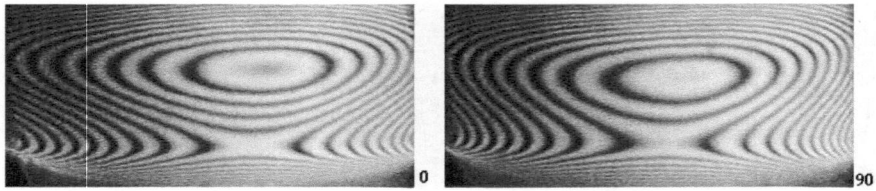

Fig. 5.17 Interferograms corresponding to $\Delta T = 2.4$ K for view angles of $0°$ (*left*) and $90°$ (*right*) in a cavity 50% filled with silicone oil

Fig. 5.18 Line-of-sight averaged temperature profile in
silicone oil for $\Delta T = 2.4$ K
and two projection angles of
Fig. 5.17

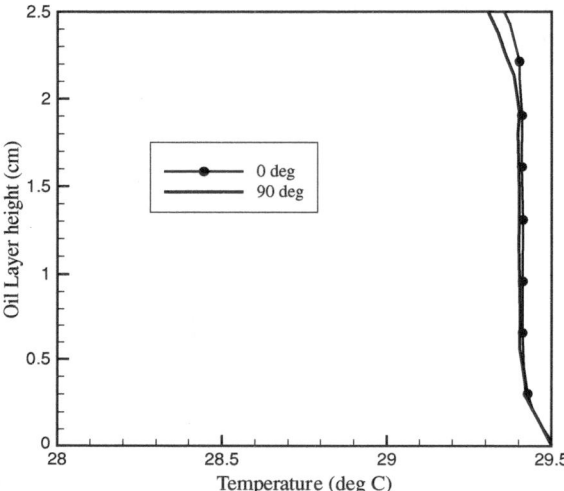

to their respective temperature differences as 2,968 and 8,881, respectively. The two
view angles of Fig. 5.17 show near-axisymmetry in the thermal fields. The vertical
temperature profiles in oil for the two projection angles are shown in Fig. 5.18.

5.4.2 Shadowgraph Analysis

The light intensity contrast, namely change in intensity with respect to the original
is related to the Laplacian of the temperature field. Assuming the temperature to
vary only in the vertical direction, we obtain the second derivative of temperature
d^2T/dy^2 at each pixel along any vertical column of the shadowgraph image. The
differential equation for temperature can be solved along with two boundary conditions. The lower wall temperature is known from experimental conditions but and
explicit information at the interface is not available a priori. As a first approach, the
energy balance equation can be used to estimate the average interface temperature.
Thus, we have the following boundary condition of the first type:

Table 5.1 Comparison of temperature and heat flux at the air–oil interface for $\Delta T = 2.4$ K when the cavity is half-filled with silicone oil

At interface	Interferometry	Shadowgraph (1)	Shadowgraph (2)	Shadowgraph (3)	Based on energy balance
Temperature (°C)	29.20	29.11	29.13	28.88	29.11
Heat flux (W/m²)	30.80	15.87	20.93	32.39	20.93

$$\text{lower wall } T = T_{\text{hot}}$$
$$\text{interface } T = T_{\text{interface}} \qquad (5.4)$$

Here, the interfacial heat flux in silicone oil is determined using the Nusselt number correlation for the portion of the cavity containing air. Hence, at the interface we have

$$-k_{\text{oil}} \left. \frac{\partial T}{\partial y} \right|_{\text{interface}} = h_{\text{air}}(T_{\text{interface}} - T_{\text{cold}}) \qquad (5.5)$$

The heat transfer coefficient h_{air} is obtained from a single fluid correlation such as Eq. 5.2. Hence, the two boundary conditions can be summarized as:

$$\text{lower wall } T = T_{\text{hot}}$$
$$\text{interface } \left. \frac{\partial T}{\partial y} \right|_{\text{interface}} = \text{given} \qquad (5.6)$$

Alternatively, from a single fluid correlation in silicone oil, the average heat flux at the lower surface can be determined. Hence, the third set of boundary conditions are:

$$\text{lower wall } T = T_{\text{hot}}$$
$$\text{lower wall } \left. \frac{\partial T}{\partial y} \right|_{y=0} = \text{given} \qquad (5.7)$$

In the following discussion, Eqs. 5.4, 5.6, and 5.7 are identified as *Shadowgraph-1*, *Shadowgraph-2*, and *Shadowgraph-3*, respectively. In the present experiment, shadowgraph images have been recorded in the interferometer itself by blocking the reference beam. The comparison between the predictions of temperature and interfacial heat flux by the two optical techniques is given in Table 5.1. The comparison of shadowgraph analysis with interferometry for a temperature difference of 2.4 K across the cavity is shown in Fig. 5.19. Three layer heights of silicone oil are considered. Image analysis using Shadowgraph-1 (Eq. 5.4) shows the best match with interferograms. Interferometry and shadowgraph match identically at the lower wall where the wall temperature is prescribed in the analysis of both images. The fact that Eq. 5.4 predicts a temperature at the interface close to that of interferometry

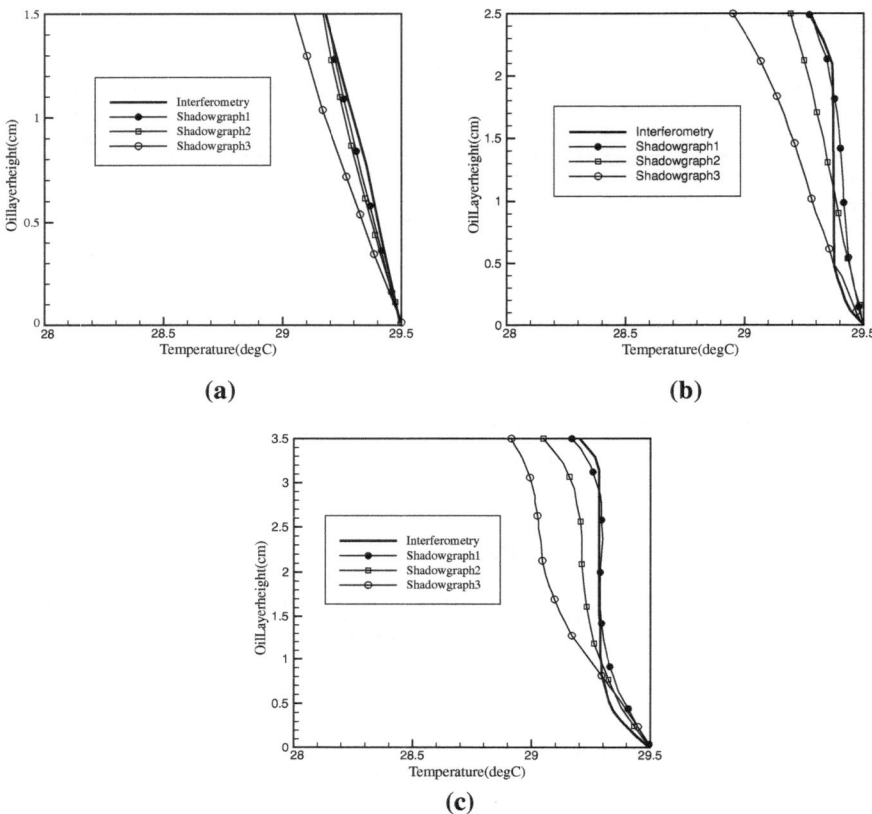

Fig. 5.19 Comparison of temperature profiles determined using interferometry and shadowgraph analysis for $\Delta T = 2.4$ K in a cavity partially filled with silicone oil; **a** 30, **b** 50, and **c** 70 %

validates the shadowgraph analysis procedure. The overall success of Eq. 5.4 is also to be expected since the interferograms indeed show the interface to be nearly an isotherm. There are, however, differences in terms of temperature gradients, particularly near the wall and the interface. Here, interferometry is possibly inaccurate owing to refraction errors and difficulty in precisely locating the free surface boundary. The predictions of Shadowgraph-2 (Eq. 5.6) are close to interferometry for the 30 % filled cavity. The differences increase sharply at other oil heights. Boundary conditions given by Shadowgraph-3 (Eq. 5.7) show the largest deviation with respect to interferometry, both in terms of gradients as well as temperature. This is expected because both boundary conditions are applied at the same location, a numerically ill-posed formulation. Based on the discussion above, shadowgraph images have been evaluated using the 1D form of Laplace equation and boundary conditions given by Eq. 5.4.

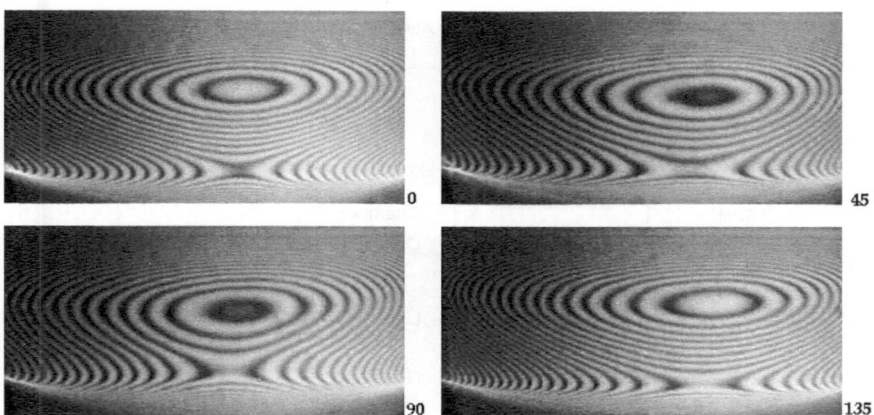

Fig. 5.20 Interferograms at $\Delta T = 5\,\mathrm{K}$ for projection angles 0, 45, 90, and 135° for cavity 50% filled with silicone oil

5.5 Buoyant Jets

The measurement of concentration distribution within a buoyant jet has significance in various fuel–air mixing studies and fire applications [4]. The schematic drawing of the experimental apparatus is shown in Fig. 5.21. The nozzle is located inside a 340 mm long test cell. The cross-section of the test cell is octagonal with each side equal to 8.5 mm. The dimension of the test cell is larger than the nozzle (diameter = 5 mm). It is expected that wall effects are small and a free jet behavior is realized. The test cell is quite long so that outflow influences do not disturb flow in the near-field of the nozzle. In addition, a wire mesh on the outflow plane breaks up the jet and minimizes the possibility of large-scale structures formed within the text cell. Atmospheric air enters the test cell through a honeycomb section and a wire mesh. The nozzle is positioned at the center of the octagonal test cell. The nozzle is 32 cm long with a length-to-diameter ratio equal to 64. The flow at the exit plane of the nozzle is expected to be laminar and fully developed. The nozzle is connected to a helium gas cylinder through a special-purpose rotameter (*Scientific Devices Pvt. Ltd.*) and a flow control valve.

Figure 5.22 shows a color schlieren pattern recorded when a helium jet mixes with the surrounding air. From the local density of the mixture, the mole faction of air (and hence, oxygen) can be worked out. Figure 5.23 compares the oxygen concentration distribution in the helium jet with the data of [1]. The concentration profiles are similar in the two studies with a maximum deviation of 10%. This deviation may be attributed to the error in positioning of the jet. Figure 5.23 also confirms successful implementation of the color schlieren analysis technique for determination of the concentration distribution.

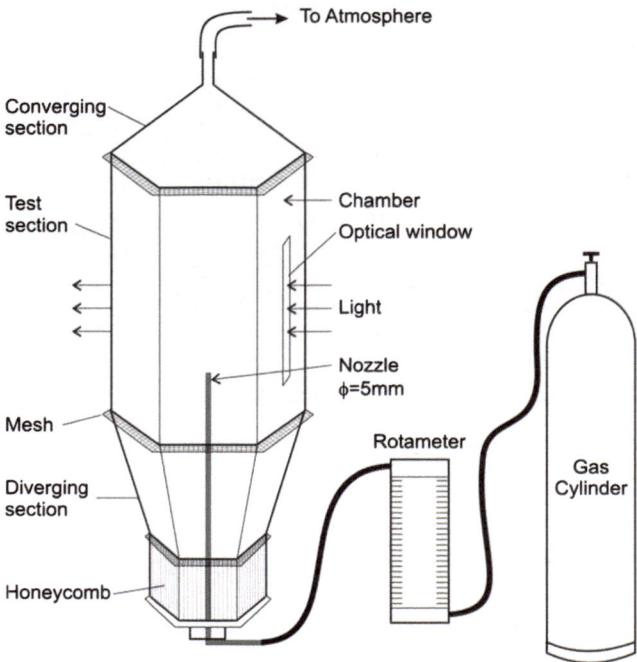

Fig. 5.21 Schematic drawing of the experimental setup for a buoyant jet experiment

Fig. 5.22 Color schlieren image of a vertically oriented buoyant helium jet mixing with air under ambient conditions

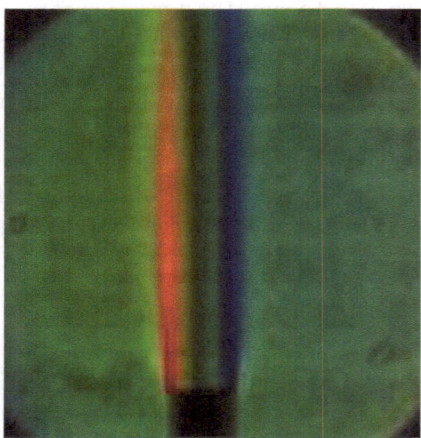

5.6 Wake of a Heated Cylinder

The wake of a bluff body is unsteady, 3D, and undergoes transitions with increase in Reynolds number [8]. When the body is heated, buoyancy forces can alter the

Fig. 5.23 Comparison of the oxygen mole fraction determined from the color schlieren image (Fig. 5.22) of the present study with [1]

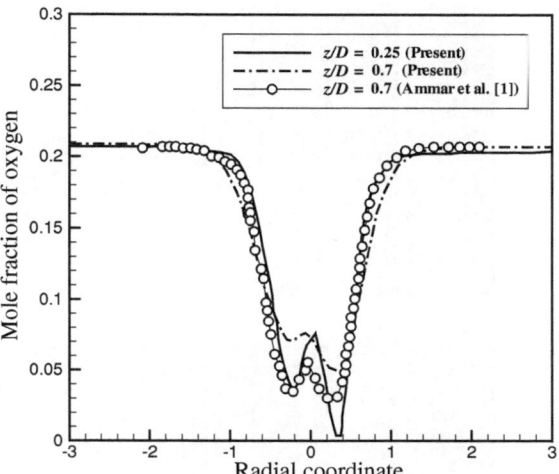

flow patterns. In addition, changes in fluid viscosity can also play an important role. Unsteady behavior of the wake of a heated bluff body has immense significance for heat exchanger and flow control applications.

The experimental apparatus and the schlieren configuration used for imaging the wake of fixed and moving cylinders of square and circular cross-sections are discussed in Chap. 6. The present discussion is for a square cylinder of edge B. The main flow is in the vertically upward direction while the cylinder axis is horizontal. The incoming flow is cold with respect to the cylinder. Here, the definition of Reynolds number is based on the average incoming fluid speed (U), the cylinder edge B, and fluid properties based on the inflow fluid temperature, around 24 °C. In addition, the Richardson number Ri is defined as

$$\text{Ri} = \frac{g\beta\Delta T B}{U^2} \tag{5.8}$$

Figure 5.24 shows instantaneous schlieren images as a function of Richardson number for flow past a heated square cylinder at Re $= 87$. The initial schlieren image before the start of the experiment is a dark patch. Streaks of light are visible only when the cylinder is heated relative to the incoming fluid. When the heating level is low, these images show alternate shedding of vortices from the opposite sides of the wake centerline. Mixed convection effects are visible at higher heating levels. For Ri ≥ 0.171, vortex shedding was seen to be suppressed and a steady plume was obtained.

Similar experiments were conducted with a heated horizontal circular cylinder, with the main flow in upward direction. The critical Richardson number at which vortex shedding was suppressed was monitored. These values are compared against the published literature in Chap. 6 and comparison was found to be quite favorable.

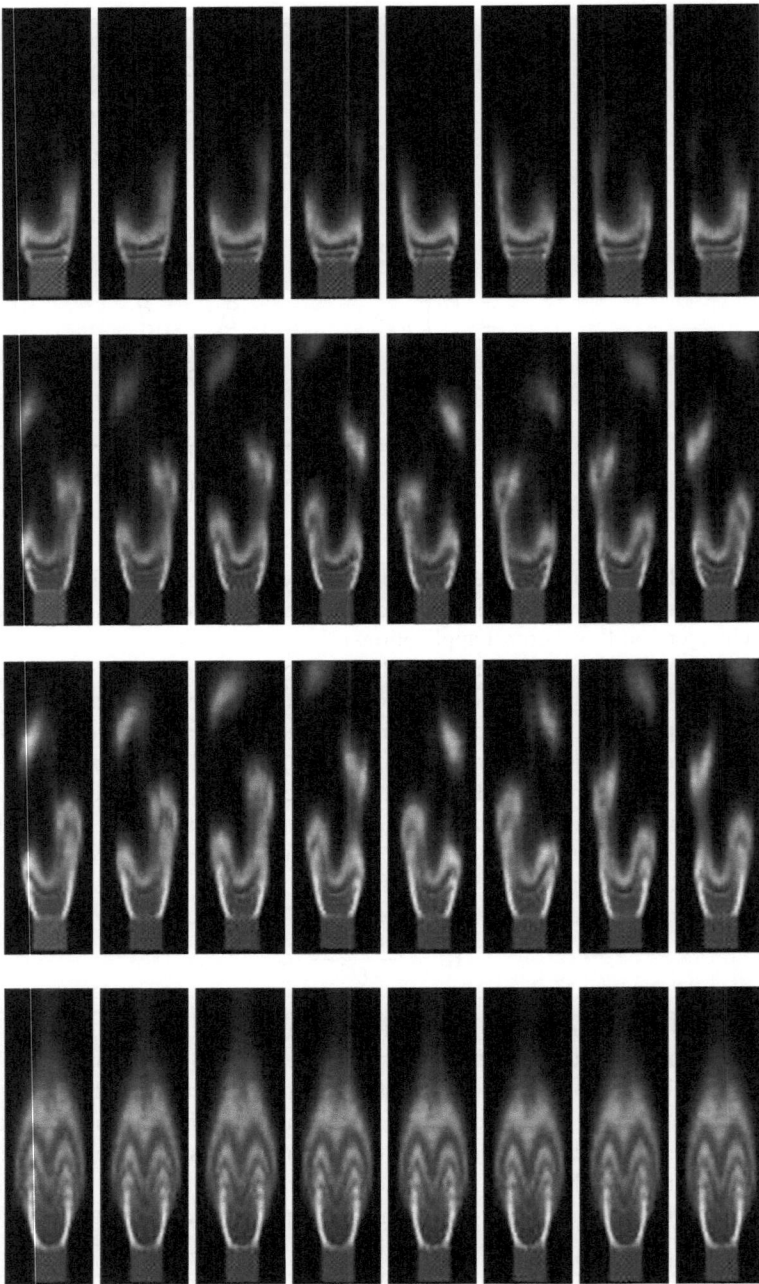

Fig. 5.24 Instantaneous schlieren images for vertically upward flow past a heated square cylinder; images are separated by a time interval of one eighth of the time period of vortex shedding. Reynolds number = 103. Row-wise, Ri = 0.044, 0.079, 0.108 and (last row) 0.138. For Ri > 0.138, images show a steady plume and vortex shedding is suppressed

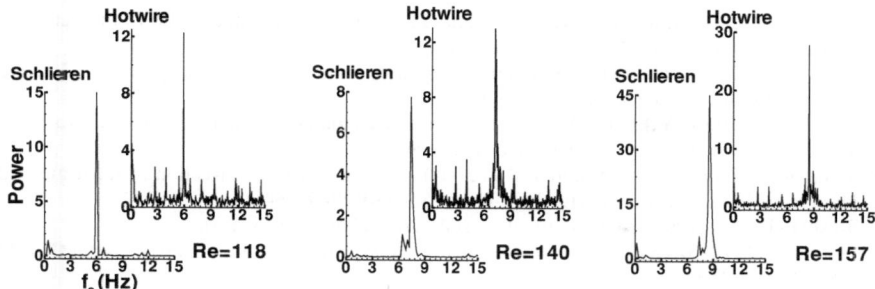

Fig. 5.25 Comparison of power spectra of the hotwire and schlieren signal in the wake of a heated square cylinder

Comparison of other wake parameters and consistency checks in temperature measurements are also discussed in Chap. 6.

5.6.1 Comparison of Hotwire and Schlieren Signals

Since optical imaging is interia-free, a high-speed camera can record a large number of frames of the wake patterns. At each pixel of the frame, light intensity is available and a time series can be constructed. In this respect, temperature change in time serves as a thermal tracer that manifests as light intensity fluctuations. In gases, where the Prandtl number is around unity, these fluctuations are a measure of velocity fluctuations as well.

Here, spectra of light intensity fluctuations in the wake of a heated square cylinder are compared with those measured using a hotwire anemometer. The location selected is in the intermediate wake so that hotwire signals are not damped by an overall increase in the bulk fluid temperature. Figure 5.25 shows a comparison of the power spectra of the hotwire probe and those calculated from schlieren signals at three Reynolds numbers. It can be seen that the spectral peak occurs at nearly equal frequencies in the two measurements. The light intensity spectra are sharper and less noisy. The advantage stems from the fact that optical techniques are non-intrusive. Thus, schlieren can be recommended for optical imaging of wake structures of bluff objects.

References

1. Al-Ammar K, Agrawal AK, Gollahalli SR, Griffin D (1998) Application of rainbow schlieren deflectometry for concentration measurements in an axisymmetric helium jet. Exp Fluids 25:89–95

2. Gebhart B, Jaluria Y, Mahajan RL, Sammakia B (1988) Buoyancy-induced flows and transport. Hemisphere Publishing Corporation, New York
3. Narayanan R, Schwabe D (2003) Interfacial fluid dynamics and transport processes. Lecture notes in physics. Springer, Berlin
4. Phipps MR, Jaluria Y, Eklund T (1997) Helium-based simulation of smoke spread due to fire in enclosed spaces. J Combust Sci Technol 157:6386
5. Punjabi S, Muralidhar K, Panigrahi PK (2004) Buoyancy-driven convection in superimposed fluid laters in an octagonal cavity. Int J Therm Sci 43(9):849–864
6. Rashkovich LN (1991) KDP family of crystals. Adam Hilger, New York
7. Wilcox WR (1993) Transport phenomena in crystal growth from solution. Prog Crystal Growth Char 26:153–194
8. Williamson CHK (1996) Vortex dynamics in the cylinder wake. Annu Rev Fluid Mech 28: 477–539

Chapter 6
Closure

Keywords Comparison · Interferometry · Schlieren and shadowgraph · Flow visualization · Ease of analysis

6.1 Introduction

Imaging convection patterns in transparent media using variations in refractive index are discussed in this monograph. The images are analyzed to obtain temperature and solutal concentration distributions. While the schlieren technique is emphasized, related methods such as interferometry and shadowgraph are also introduced. These methods have been validated against data available in the literature. The information contained in optical images is projection of the field variable, such as temperature, in the sense that it is *path integral* in the viewing direction. The local distribution of temperature (or concentration) can be retrieved by an analytical procedure called tomography. Under certain conditions, this approach can be used for time-dependent fields as well. The present chapter compares the three optical techniques in terms of ease of setting up and data analysis.

6.2 Comparison of Interferometry, Schlieren, and Shadowgraph

Images recorded from interferometry, schlieren, and shadowgraph yield the field variable, its first derivative, and the Laplacian, respectively. Accordingly, schlieren and shadowgraph data have to be integrated to recover quantities such as temperature and concentration. It may appear as if interferometry is superior to schlieren and shadowgraph since it requires minimal analysis. However, other factors may be pertinent as listed below and will determine the best approach in a given context.

P. K. Panigrahi and K. Muralidhar, *Schlieren and Shadowgraph Methods in Heat and Mass Transfer*, SpringerBriefs in Thermal Engineering and Applied Science, DOI: 10.1007/978-1-4614-4535-7_6, © The Author(s) 2012

1. Interferometry relies on differential phase measurement using two light beams and requires additional optical elements for generating interferograms. Schlieren and shadowgraph work with a single test beam and considerably simplified optical layout. Schlieren requires a decollimating arrangement of the refracted light beam. Shadowgraph is the simplest of the three.

2. Interferogram analysis depends on locations of intensity minima but not the intensity itself but is easily contaminated by refraction errors. Schlieren and shadowgraph data rely on light intensity measurement and can be affected by camera linearity and saturation. Diffraction at the knife-edge and errors in gray-scale and color filters are also relevant in schlieren. If higher order effects are significant, shadowgraph analysis can become numerically intractable and the images may have only qualitative utility.

3. In experiments with low density gradients, interferograms are clear and useful but intensity contrast in schlieren and shadowgraph may not be large enough to provide a vivid picture. In high gradient experiments, both schlieren and shadowgraph will yield clear and interpretable images. Interferograms are, however, corrupted by refraction errors.

4. In unsteady flow fields, schlieren and shadowgraph will track temporal changes in temperature and concentration in the form of light intensity variations. These are less obvious in interferograms where information is localized at fringes.

5. The number of fringes formed in liquids such as water and silicone oil is quite large since the sensitivity of refractive index to density $(dn/d\rho)$ is large in these media. Interferograms recorded in such experiments are affected by refraction errors. Schlieren and shadowgraph hold an advantage in this respect.

6. In a flow field dominated by vortices and other length and timescales, schlieren and shadowgraph offer the advantage of clarity, not readily derivable from interferometry.

Overall, the simplicity of analysis, ease of instrumentation, and adaptability for a wide range of applications make schlieren a preferred choice.

6.3 Applications

A companion volume by the authors entitled *Imaging Heat and Mass Transfer Processes—Visualization and Analysis* is concerned with refractive index-based imaging in a variety of applications. These include

1. growth of a crystal from its aqueous solution,
2. flow past heated bluff bodies,
3. Interfacial transport in superposed fluid layers, and
4. buoyant jets.